Mental Health in Historical Perspective

Series Editors
Catharine Coleborne
School of Humanities and Social Science
University of Newcastle
Callaghan, NSW, Australia

Matthew Smith
Centre for the Social History of Health and Healthcare
University of Strathclyde
Glasgow, UK

Covering all historical periods and geographical contexts, the series explores how mental illness has been understood, experienced, diagnosed, treated and contested. It will publish works that engage actively with contemporary debates related to mental health and, as such, will be of interest not only to historians, but also mental health professionals, patients and policy makers. With its focus on mental health, rather than just psychiatry, the series will endeavour to provide more patient-centred histories. Although this has long been an aim of health historians, it has not been realised, and this series aims to change that.

The scope of the series is kept as broad as possible to attract good quality proposals about all aspects of the history of mental health from all periods. The series emphasises interdisciplinary approaches to the field of study, and encourages short titles, longer works, collections, and titles which stretch the boundaries of academic publishing in new ways.

More information about this series at
http://www.palgrave.com/gp/series/14806

Hannah Proctor

Psychologies in Revolution

Alexander Luria's 'Romantic Science' and Soviet Social History

palgrave
macmillan

Hannah Proctor
Centre for the Social History of Health and Healthcare
University of Strathclyde
Glasgow, UK

Mental Health in Historical Perspective
ISBN 978-3-030-35027-7 ISBN 978-3-030-35028-4 (eBook)
https://doi.org/10.1007/978-3-030-35028-4

Cover illustration: Book Worm / Alamy Stock Photo

This Palgrave Macmillan imprint is published by the registered company Springer Nature Switzerland AG.
The registered company address is: Gewerbestrasse 11, 6330 Cham, Switzerland

Acknowledgements

Laura Salisbury is responsible for introducing me to Alexander Luria's work. She supported this project from the outset and encouraged me throughout. Esther Leslie gave me invaluable advice on the structure of my arguments and its theoretical framing. Comments from Howard Caygill and Daniel Pick helped me to reshape and reconceptualise the project for publication, as did those of an anonymous peer reviewer of the manuscript.

A doctoral award from the Arts and Humanities Research Council (AHRC) made the PhD out of which this book grew possible. Thanks to Mary Lou Reker and the staff at the John W. Kluge Center at the Library of Congress for all their help during my fellowship there (also funded by the AHRC). Thanks to the archivists who helped me to locate materials at The Institute of Education (London), Wellcome Collection (London), The American Philosophical Society (Philadelphia), Northwestern University Library Archives (Chicago), American Jewish Archives (Cincinnati), Harvard University Archives (Cambridge, MA) and Johns Hopkins Archives (Baltimore). Special thanks to Lizette Royer Barton at the Drs. Nicholas and Dorothy Cummings Center for the History of Psychology, University of Akron (Ohio); Tatiana Anokhina at RGALI (Russian State Archive of Literature and Art) (Moscow); and—at the final hour!—Kate Edgar who ensured I was able to access correspondence between Luria and Oliver Sacks. I'm grateful to the late Jerome Bruner for taking time to talk to me whilst I was in New York in 2014, to Mike Cole for sharing his personal correspondence with Luria and to Anton Yasnitsky for making reading recommendations (particularly in relation to the

Central Asian expeditions way back when this project was just beginning to take shape).

Thanks to everyone at the ICI Berlin where this project began its transition from thesis to book—Christoph Holzhey, Manuele Gragnolati, Claudia Peppel, Arnd Wedermeier as well as Frauke Both, Chris Wunsch and my cohort of fellows.

Elena Khrianina helped me with so much in Moscow where I was also lucky to first meet Maria Chehonadskih. I think fondly of the late Olga Sharapova who took me for peaches and smoked Vogue cigarettes in a rainstorm at Chistye Prudy. A more recent invitation from Keti Chukrov provided the impetus to track down the Sacks correspondence. Thanks to my parents Marilyn and Neil Proctor and to the friends and housemates who read my work or talked to me about different aspects of it over the years: Larne Abse Gogarty, Jacob Bard Rosenberg, Hannah Black, Richard Braude, Christina Chalmers, Anna de Filippi, Sam Dolbear, Sam Goff, Rose-Anne Gush, Owen Hatherley, Sophie Jones, Ed Luker, Michael Runyan, Amy Tobin, Daniel Reeve, Andrew Witt and many many others I've probably failed to mention.

Special thanks to the people who came to visit me in hospital in Berlin when I had hoped to be in the very final stages of writing: Sarah Crane Brewer, Lizzie Homersham, Max King, Eve Richens and Michelle Ty (and Daniel Reeve, Michael Runyan and Andrew Witt again).

Thanks also to Catharine Coleborne and Matthew Smith, the editors of the series Mental Health in Historical Perspective, and to Molly Beck and Maeve Sinnott at Palgrave Macmillan.

The genre dictates that one thanks one's family or partner last, but this thesis owes as much to collaborations, conversations and collective projects as it does to any particular individuals. Sharing ideas with friends while living, researching and writing in London, Moscow, Washington D.C. and Berlin, however fleeting those encounters often were, made completing this feel like less of an isolated endeavour.

Note on Transliteration

Except in the case of proper names familiar to the English-language reader (e.g. Luria, Trotsky, Vygotsky and Krupskaya), Russian words have been transliterated according to the Library of Congress scheme and are italicised in the text.

Contents

CHAPTER 1

Introduction

'Let us assume a brain has been removed from its cranium and placed on a glass table before us.'[1] In a discussion of the localisation of brain function, Alexander Luria (1902–1977) invited his readers to imagine a brain removed from the body and displayed in isolation, its gelatinous surface visible from all angles. In such a situation, he explained, it would be possible to observe a fleshy grey mass of tissue ridged with 'deep furrows and raised convulsions'.[2] The 'uniform and monotonous' appearance of this lump of dead meat, however, belies the living brain's extraordinary complexity and dynamism.[3] Observing a brain in such a manner, Luria suggested, tells us very little about human experience. His solution was to reconnect brains with people and to situate those people in the world. As he commented in a lecture series delivered in 1976, towards the end of his life:

> In order to explain the highly complex forms of human consciousness one must go beyond the human organism. One must seek the origins of conscious activity and 'categorical' behaviour not in the recesses of the human brain or in the depths of the spirit, but in the external conditions of life.

[1] Alexander Luria, *The Man with a Shattered World: The History of a Brain Wound*, trans. by Lynn Solotaroff (Cambridge, MA: Harvard University Press, 2002), p. 22. First published in Russian as *Poteriannyi i vozrashchennyi mir: Istoriia odnogo raneniia* [*A World Lost and Re-gained: The Story of an Injury*] (Moscow: Izdatel'stvo MGU, 1971).

[2] Luria, *Shattered*, p. 22.

[3] Luria, *Shattered*, p. 23.

H. Proctor, *Psychologies in Revolution*, Mental Health in Historical Perspective, https://doi.org/10.1007/978-3-030-35028-4_1

> Above all, this means that one must seek these origins in the external processes of social life, in the social and historical forms of human existence.[4]

Writing almost four decades earlier, he described his aims in strikingly similar terms:

> Modern psychology has come to the firm view that human personality is shaped by its concrete sociohistorical circumstances. We can think of no form of behaviour that can be studied in isolation from this historical context, by itself, independent of the specific sociohistorical conditions determining it.[5]

Luria consistently asserted that human consciousness could not be understood in isolation from history, culture or society.

Born in Kazan in 1902, Luria trained and began to work as a psychologist in the aftermath of the October Revolution of 1917. He moved to Moscow in 1923 where, aside from a brief spell in the Ukrainian city of Kharkiv in the early 1930s, he lived and worked for the rest of his life. He received a medical qualification in 1937 after which his research extended to encompass neurological investigations. Reading lists of institutions, disciplines and organisations with which Luria was associated over the course of his long working life indicates that his was a career of almost bewildering diversity which saw him engage in fields across and occasionally beyond the psy-ences[6]—from psychoanalysis to criminology, neurosurgery to 'defectology', experimental medicine to pedagogy. Luria's bibliography,

[4] Luria, *Language and Cognition*, trans. by James V. Wertsch (Washington, DC: VH Winston and Sons, 1981), p. 25.

[5] Luria, 'A Child's Speech Responses and the Social Environment' in *Soviet Developmental Psychology*, ed. by Michael Cole (New York, NY: M. E. Sharpe, 1977), pp. 32–64, p. 35, p. 60. Excerpted from the edited collection *Speech and Intellect among Rural, Urban and Homeless Children* [*Rech' i intellekt derevenskogo, gorodskogo i besprizornogo rebenka*] (Moscow: Gosizdat RSFSR, 1930), p. 32.

[6] The term 'psy-ences' was introduced by Elizabeth Lunbeck, Emily Martin and Louis Sass in the description of a seminar series and has since been applied to the Soviet and post-Soviet context in Eugene Raikhel and Dörte Bemme, 'Postsocialism, the psy-ences and mental health', *Transcultural Psychiatry*, 53, 2 (2016), 151–175 and Tomas Matza, *Shock Therapy: Psychology, Precarity, and Well-Being in Postsocialist Russia* (Durham, NC: Duke University Press, 2018). Raikhel and Bemme define the term as referring 'broadly to a set of arguments which link professional knowledge and expertise on the mind, brain, and behavior with the wide range of ways in which people conceive of, act upon, and govern themselves and others under conditions of modernity', p. 154.

which runs to approximately 350 publications in Russian alone, is similarly expansive.[7] His writings encompass monographs on handwriting, memory, speech and children's play; theoretical articles on the relationship of Marxism to psychoanalysis; diagram-heavy textbooks on the localisation of functions in the cerebral cortex; batteries of tests for use by clinicians; and two case histories written in a literary style for a mass audience. His clinical work brought him into contact with a correspondingly broad range of test subjects and patients, both 'normal' and pathological. Luria's shifts in disciplinary focus and institutional affiliation were partly dictated by the shifting priorities of the Soviet state, which forced him to abandon certain disciplinary approaches at particular moments. Yet in spite of these external exigencies and the undeniably capacious scope of his expertise, Luria's central concerns remained remarkably consistent across his long career. Indeed, he resisted drawing any clear distinctions between his work in such seemingly discrete disciplinary spheres at all. As his biographer and erstwhile collaborator Evgenia Homskaya notes:

> All his life he worked at the junction of several different sciences. He always saw the subject of his study in its entirety (as a 'whole') and was able to synthesise fragmentary knowledge into a harmonious system.[8]

He coined the term 'neuropsychology' to describe the new 'synthetic' scientific discipline he sought to create.[9]

Luria dedicated time and energy to establishing conversations with psychologists, educators and neurologists in the West, ensuring that his work reached an international audience beyond the Eastern bloc. K.E. Levitin recalled that Luria's apartment had a 'huge custom-made mailbox' to

It is an extension of the term 'psy disciplines', which Nikolas Rose introduced to designate a range of scientific practices that since the nineteenth century have participated in determining how human beings understand themselves and that have helped to enable new forms of governance. See, Nikolas Rose, *Inventing Our Selves: Psychology, Power and Personhood* (Cambridge: Cambridge University Press, 1998), pp. 1–21. Rose briefly discusses the Soviet psy-ences in this publication, declaring that much about the subject 'remains to be analysed', p. 15.

[7] For a bibliography of Luria's publications see: Evgenia D. Homskaya, *Alexander Romanovich Luria: A Scientific Biography*, trans. by Daria Krotova (New York, NY: Plenum Press, 2001), pp. 127–161.

[8] Homskaya, pp. 1–2.

[9] See A.R. Luria, *The Working Brain: An Introduction to Neuropsychology*, trans. by Basil Haigh (New York, NY: Basic Books, 1973), p. 105.

accommodate his voluminous correspondence.[10] Luria's position in 'world science' is assured; his publications still cited and revered.[11] His mid-career shift from psychoanalytic and psychological research into neurological investigations reflects broader developments in the psy-ences across the twentieth century, in dialogue with yet divergent from his contemporaries on the other side of the iron curtain. Aside from his autobiography and biographies (written mostly by former colleagues and a family member),[12] almost all of the existing literature on Luria is written by and for practicing psychologists, neurologists and educators.[13] Yet Luria himself emphasised the importance of situating theories in history noting that 'the eye of science does not probe "a thing", an event isolated from other things or events. Its real object is to see and understand the way a thing or event relates to other things or events.'[14] *Psychologies in Revolution* intends to apply this insight to Luria's own publications, contending that analysing Luria's research in isolation from the historical circumstances it emerged from and influenced would be like analysing someone's personality by examining their brain on a glass table.

[10] See K.E. Levitin, 'Epilogue: Luria's Psychological Symphony', *Journal of Russian and East European Psychology*, 6, 36 (1998), 33–62, p. 53. Luria's former collaborator Michael Cole informed me that Luria set aside a few hours every day to keep up with his international correspondence (telephone interview, 23 September 2014). Similarly Oliver Sacks recalled that 'Luria, after a twelve or sixteen hour working day, would spend hours more with an enormous scientific correspondence, writing constantly to colleagues, former pupils, and friends, detailed, passionate letters, in half a dozen different languages'. Oliver Sacks, 'Luria and "Romantic Science"' in *The Cambridge Handbook of Cultural-Historical Psychology*, ed. by Anton Yasnitsky, René van der Veer and Michel Ferrari (Cambridge: Cambridge University Press, 2014), pp. 517–528.

[11] For overviews of Luria's global influence, see: R.L. Solso and C.A. Hoffman, 'Influence of Soviet Scholars', *American Psychologist*, 46 (1991), 251–253 and David E. Tupper, 'Introduction: Alexander Luria's Continuing Influence on Worldwide Neuropsychology', *Neuropsychology Review*, 9 (1999), 1–7.

[12] T.V. Akhutina, 'A.R. Luriia: zhiznennyi put' [A.R. Luria: The Way of Life], *Kul'turno-istoricheskaya psikhologiia*, 2 (2012), 2–10, Evgenia D. Homskaya, *Alexander Romanovich Luria: A Scientific Biography*, trans. by Daria Krotova (New York, NY: Plenum Press, 2001), K.E. Levitin, *Mimoletnyi uzor* [*A Transient Pattern*] (Moscow: Izd-vo, 1978) (serialised in English translation in the *Journal of Russian and East European Psychology* in 1998) and E. Luria, *Moi Otets A.R. Luriia* [*My Father A.R. Luria*] (Moscow: Gnosis, 1994).

[13] See, for example, Anne-Lise Christensen, Elkhonon Goldberg and Dmitri Bougakov eds., *Luria's Legacy in the 21st Century* (Oxford: Oxford University Press, 2009).

[14] Luria, *The Making of Mind: A Personal Account of Soviet Psychology*, ed. by Michael Cole and Sheila Cole (Cambridge, MA: Harvard University Press, 1979), p. 174. This book was written directly in English and subsequently translated into Russian.

In 1976, a year before his death, Luria wrote to his old friend the American psychologist Jerome Bruner, to inform him that he was working on a 'highly personal' and possibly final book, declaring his intention 'not to write an autobiography but rather a history of a social atmosphere after the revolution with all the enthusiasm of trying to find new ways'.[15] The resulting text, *The Making of Mind*, written directly in English and published posthumously in 1979, forgoes a description of Luria's childhood experiences and begins instead in 1917 with the October Revolution (by which time he was already 15 years old):

> I began my career in the first years of the great Russian Revolution. This single, momentous event decisively influenced my life and that of everyone I knew. ... My entire generation was infused with the energy of revolutionary change—the liberating energy people feel when they are part of a society that is able to make tremendous progress in a very short time.[16]

Luria clearly aligned his own personal and professional development with the fate of the Soviet project with which his adult life was almost coterminous (he died in 1977, less than a decade before the introduction of Mikhail Gorbachev's sweeping reforms).[17] But as he noted in the autobiography's conclusion: 'People come and go, but the creative sources of great historical events and the important ideas and deeds remain.'[18]

Psychologies in Revolution eschews a detailed biographical approach to Luria as an individual, makes no attempt to describe his temperament or emotional sensibilities, says almost nothing about his family life and little about the many people with whom he collaborated professionally. Instead, it seeks to bring the 'social atmosphere' in which he conducted his research into sharper focus. Beginning in the aftermath of the October Revolution when Luria actively engaged in attempts to conceptualise what a properly

[15] Alexander Luria to Jerome Bruner, March 16, 1976, HUA, Jerome Bruner Papers, General Correspondence 1975–1977, HUG 4242.5, Box 88. Luria had written to Michael Cole shortly before writing to Bruner that the working title for his autobiography was: 'The Last Book'. Luria to Cole, March 2, 1976, Michael Cole Personal Archives.

[16] Luria, *Making of Mind*, p. 17.

[17] For an analysis of his early career, see Michael Paul George Hames, 'The Early Theoretical Development of Alexander Luria', PhD, University of London, UCL, 2002. A concise overview of Luria's career is given in: M.I. Kostyanaya and P. Rossouw, 'Alexander Luria—Life, Research and Contribution to Neuroscience', *International Journal of Neuropsychotherapy*, 1, 2 (2013), 47–55.

[18] Luria, *Making of Mind*, p. 188.

Marxist approach to human consciousness might entail, through the period of the First Five Year Plan (1928–1932) when he participated in projects seeking to gauge the cognitive impact of Stalinist policies, to the Second World War which saw him turn his attention to the rehabilitation of brain-injured soldiers, *Psychologies in Revolution* focuses on moments in Luria's career that coincided with particular moments in Soviet history, considering how both his research and the people it brought him into contact with were shaped by that history. Pushing Alexander Luria the man into the background also allows for the people he encountered in his research to come more clearly into view.[19] *Psychologies in Revolution* is less interested in the 'extraordinary person'[20] behind the name embossed on the spine of Luria's books than in the often marginalised Soviet people Luria and his collaborators sought to understand, describe, diagnose or treat.[21]

The Science of Social History

In February 1927, a basket appeared at a train station in Moscow addressed to the city of Briansk. Suspicious about its contents, workers at the station opened the basket to discover it contained a corpse. Dressed in a tunic, wrapped in paper and tied up with ropes, a woman's dead body was squashed into the basket. She had evidently been killed with a blunt instrument. The paper around her body was found to be a disposable tablecloth,

[19] Though the book does not discuss interpersonal relationships between the individuals with whom he worked in any detail, by de-emphasising Luria as an individual I hope implicitly to acknowledge the collaborative nature of his research, which saw him devising and conducting experiments collectively, even though his publications tended to be sole-authored. Treating Luria as a solitary genius also risks downplaying one of the most distinctive elements of Soviet psychology: the prominent role allotted to women. Collaboration is the main lens through which Anna Stetsenko has interpreted Luria and Vygotsky's working practices. See, for example, Anna Stetsenko and Ivan Arievitch, 'Vygotskian Collaborative Project of Social Transformation: History, Politics, and Practice in Knowledge Construction', *The International Journal of Critical Psychology*, 12, 4 (2004), 58–80.

[20] Homskaya, p. 1.

[21] Though people Luria conducted experiments with may have ended up being classified as such and often belonged to demographics who routinely were, the term 'marginalised' here does not intend to imply official forms of disenfranchisement discussed by Golfo Alexopoulos, 'the millions who were deprived of their rights in the Stalin era and known as the *lishentsy*, a Russian word which literally means "the deprived" but is often translated as "the disenfranchised"'. Golfo Alexopoulos, *Stalin's Outcasts: Aliens, Citizens, and the Soviet State, 1926–1936* (Ithaca, NY: Cornell University Press, 2003), p. 1.

which revealed where the murder had taken place: a restaurant called 'The Bear'. As this gruesome scene unfolded in Moscow, a man in Kyiv received a letter containing a baggage check coupon. The anonymous letter which accompanied it informed him that he could collect his wife in Briansk, a city approximately five hours by train from Moscow, explaining that she could not be sent all the way to Kyiv as the letter's sender did not have sufficient funds. The man in Kyiv was subsequently summoned to Moscow and identified his wife's body in the basket.[22] In the wake of this strange incident seven people were arrested; none of them were told of what crime they were suspected of committing. Before undergoing questioning by police the suspects were taken directly to a special laboratory where they were given a series of word association tests by a team of psychologists led by Luria, in an attempt to establish who had committed the murder. Although his work with criminals was short-lived, Luria's published work contains numerous examples of particular, peculiar and sometimes distressing glimpses into the lives and experiences of Soviet people from a range of social backgrounds—including industrial workers, rural school children, deaf babies, Uzbek *kolkhoz* members, orphaned teenagers and brain-injured Red Army soldiers—whose desires, thought processes, proclivities, memories and cognitive capacities he attempted to develop methods for understanding and treating.

Luria described psychology as the 'science of social history'.[23] As in the example of the account of the murdered body in the basket, however, Luria often contradicted his own theoretical pronouncements regarding the historical contingency of subjectivity by presenting short matter-of-fact vignettes relating to specific situations in isolation from the life trajectories and social backgrounds of the people involved. Although Luria recognised the primacy of history for understanding human subjectivity in theory, his work often failed to explore the specific impact of historical experiences on individuals in practice. As the first chapter of this book shows, in his experiments with suspected murderers, Luria was more interested in asking whether someone had committed a crime than in discovering the psychic motivations and consequences of doing so, but the

[22] Luria, *The Nature of Human Conflicts or Emotion, Conflict and Will: An Objective Study of Disorganisation and Control of Human Behaviour*, trans. by W. Horsley Gantt (New York, NY: Liveright, 1932), p. 109.

[23] A.R. Luria, *Cognitive Development: Its Cultural and Social Foundations*, trans. by Martin Lopez-Morillas and Lynn Solotaroff (Cambridge, MA: Harvard University Press, 1976), p. 6.

thumbnail cases included in that series nonetheless provide glimpses of the Soviet quotidian in Moscow during the period of the New Economic Policy (NEP) (1921–1928) that gesture towards the broader social environment in which the actions being described unfolded. *Psychologies in Revolution* intends to situate Luria's publications in the tumultuous historical circumstances out of which they arose, asking what can be gleaned about the interplay between one particular Soviet theory of subjectivity and the subjectivities of Soviet people through an analysis of the mediated and often abbreviated portraits that appear in key publications based on Luria's extensive clinical work.

Luria's research was premised on normative assumptions about individual human development. He argued that the 'advanced' human subject was the result of various developmental trajectories: the biological evolution of the species from animal to human, the cultural development of societies from 'primitivism' to 'civilisation' and the maturation of the ('healthy') individual from baby to adult. An interest in tracing the progression from 'lower' to 'higher' forms of thought united Luria's seemingly diverse strands of work. At the apex of his mountain of development stood the 'civilised' or 'cultured' [*kul'turnye*], educated and healthy adult. Luria, however, had little to say directly about this 'advanced' figure. He could only discern its outline, as though it stood on a high mountain silhouetted in front of the glaring sun. Instead, he looked at the base in order to explain the route to the peaks. The ideal person to whom much of Luria's work was tacitly addressed was conceptualised in opposition to a cluster of figures assumed to depart from that ideal. Informed by the logic of Luria's research, each chapter of *Psychologies in Revolution* is structured around one such figure, with a particular emphasis on those whose supposed cognitive 'backwardness' was connected to their social position: the criminal, the 'primitive' (Uzbek peasants with no formal education), the child, the aphasic (brain-injured Red Army soldiers) and the synaesthete.

Psychologies in Revolution builds on an existing body of work focusing on the theories and legacies of Luria's collaborator Lev Vygotsky (1896–1934),[24] but as Luria's friend Bruner noted in a letter to the British

[24] On Vygotsky, see, for example, Alexei Kozulin, *Vygotsky: A Biography of Ideas* (Cambridge, MA: Harvard University Press, 1990), René van der Veer and Jaan Valsiner, *Understanding Vygotsky: A Quest for Synthesis* (Oxford: Blackwell, 1991). And more recently: René van der Veer and Anton Yasnitsky eds., *Revisionist Revolution in Vygotsky Studies* (London: Routledge, 2016) and Anton Yasnitsky, *Vygotsky: An Intellectual Biography* (London: Routledge, 2016). For an example of the meticulous philological approach common in analyses of Vygotsky's

educator and translator Joan Simon in 1995: 'In all this Vygotskimania Luria is almost forgotten.'[25] Luria met Vygotsky in 1924, an event which he narrated as a decisive turning point in his intellectual development and they shared ideas and devised projects together for the next decade (until the latter's death from tuberculosis at the age of 37). In a letter to Oliver Sacks in 1973, he declared: 'I am only a pupil of Lev Vygotsky—the real genius of Soviet science. ... His starting point was that complex psychological processes have not an inner (biological) origin, don't start in the depths of the brain, but are the result of a social origin.'[26] Luria consistently expressed his indebtedness to Vygotsky's theorisations of the relationship between people and their environments,[27] and he was responsible for ensuring Vygotsky's works reached a wide international audience, but he outlived his former mentor by over 40 years and hence produced a much larger and more diverse oeuvre. Furthermore, unlike in many of Vygotsky's most celebrated writings, in which eloquent discussions of Marx and Engels, Spinoza and Hegel, and Shakespeare and Mayakovsky are brought to bear on psychological debates, Luria's publications rarely strayed as far from his empirical investigations and experiments.[28] Analyses of Vygotsky's work sometimes remain within the realms of intellectual history or philosophy,[29] whereas *Psychologies in Revolution* is less

work, see E. Zavershneva, 'Issledovanie rukopisi L.S. Vygotskogo "Istoricheskii smysl psikhologicheskogo krizisa"' [An investigation of the manuscript of L.S. Vygotsky's 'Historical Meaning of the Crisis in Psychology'], *Voprosy psikhologii*, 6 (2009), 119–137.

[25] Letter from Jerome Bruner to Joan Simon, 26 January 1995. London, Institute of Education (IOE), Brian Simon Papers, DC/SIM/2/10. Only three of the 21 essays in a recent collection on 'cultural-historical psychology' are explicitly concerned with Luria's work and only one of those three considers Luria in isolation from Vygotsky. See Anton Yasnitsky, René van der Veer and Michel Ferrari eds., *The Cambridge Handbook of Cultural-historical Psychology* (Cambridge: Cambridge University Press, 2014).

[26] Luria to Sacks, July 19, 1973, Oliver Sacks Archives, Oliver Sacks Foundation, New York City (transliteration modified for consistency).

[27] See, for example, the dedication to Vygotsky in A.R. Luria, *Higher Cortical Functions in Man*, trans. by Basil Haigh (London: Tavistock Publications, 1966).

[28] See, for example, Lev Vygotsky, *The Psychology of Art*, trans. by Scripta Technica, Inc. (Cambridge, MA: MIT Press, 1971) and 'The Historical Meaning of The Crisis in Psychology' in *The Collected Works of Lev Vygotsky*, vol. 3, ed. by Robert W. Rieber and Jeffrey Wollock, trans. by René van der Veer (New York, NY: Plenum Press, 1997), pp. 233–343.

[29] See, for example, Carl Ratner and Daniele Nunes Henrique Silva eds., *Vygotsky and Marx: Towards a Marxist Psychology* (Abingdon, Oxon: Routledge, 2017) or Maria Chehonadskih, 'Soviet Epistemologies and the Materialist Ontology of Poor Life: Andrei Platonov, Alexander Bogdanov and Lev Vygotsky', PhD, Kingston University, 2017.

interested in Luria's many theoretical pronouncements about the primacy of history for understanding psychology than it is in analysing how Luria's publications also ***document*** social history, providing oblique insights into the experiences and outlooks of an array of Soviet people across six decades, through accounts of his clinical encounters.

In *Russian Psychology* (1989), David Joravsky claims that Luria's autobiography, though peppered with 'standard phrases' concerning revolutionary transformation, is characterised by 'reticence on distinctly Soviet issues', particularly the class backgrounds of the people he studied.[30] He suggests Luria cynically switched approaches again and again to avoid persecution and that his works demonstrate no meaningful engagement with Marxism. Luria reacted furiously to a review by Joravsky published in 1974 in the *New York Review of Books* that made similar arguments, proclaiming dramatically that 'it evoked a deep disappointment, perhaps even a feeling of a certain disgust' for overlooking his consistent interest in demonstrating 'that the higher psychological processes are of a social-historical origin'.[31] Despite Luria's vehement protestations *Psychologies in Revolution* intends to take seriously Joravsky's accusation that Luria's '"social-historical" approach remains a declared goal, but he does not go beyond general declarations'.[32] Rather than framing this as a question of political conviction, however, this book instead seeks to unpick what Luria meant when he talked about history and to ask what that definition included and occluded.

Somewhat confusingly Luria's research was animated by two distinct and indeed contradictory understandings of history: on the one hand, he treated individual human development as a recapitulation of civilisational development (the ontogenetic maturation from childhood to adulthood was treated as a counterpart to a phylogenetic progression from primitivism to civilisation), while on the other hand he emphasised the contingent impact of specific cultural and political experiences on individuals. The former understanding is meta-historical and imagines individual

[30] David Joravsky, *Russian Psychology: A Critical History* (Oxford: Blackwell, 1989), pp. 246–248.

[31] More amusingly he also objected to being hailed as a 'great' Soviet psychologist stating bluntly: 'My work is of average value, and no more'. Luria to Sacks, September 1, 1974, Oliver Sacks Archives, Oliver Sacks Foundation, New York City.

[32] David Joravsky, 'A Great Soviet Psychologist', *New York Review of Books*, May 16, 1974, pp. 22–25, p. 23.

development as an extrinsic miniature model which replicates historical progress on an individual scale, whereas the latter defines history as the ever-shifting environment enveloping individuals in the present. In practice, however, the progressive developmental framework that informed Luria's understanding of individual development did not always map neatly onto his understanding of history as an individual's immediate social environment, life experience and inherited culture. The abstract linearity of the former could not fully account for the messy concrete realities of the latter. *Psychologies in Revolution* discerns tensions between these two forms of history on the pages of Luria's publications, emphasising instances in which the utterances and experiences they document (however partially) challenged or complicated Luria's normative assumptions about ideal forms of cognition. By questioning Luria's insistence that human subjects should ideally develop *like* history (and that history should ideally develop teleologically through a series of stages punctuated by dialectical transformations), *Psychologies in Revolution* intends to bring the experiences of people who lived *in* history to the fore.[33]

Mapping Soviet Psychic Territories

Questions of agency and conceptualisations of self-hood are central to historiographical accounts of Soviet history, which often rely, however implicitly, on assumptions about human psychology.[34] In accounts of the

[33] Following the insights of Michel Foucault's 'Lives of Infamous Men', I have sought in Luria's scientific publications glimmers of 'lives whose disarray and relentless energy one senses beneath the stone-smooth words'. Foucault intended to assemble a collection of moments when obscure people suddenly seemed to leap out from dry archival documents whose historical function was to control them; he hoped to register moments of textual dissonance where official language was disturbed, ruffled or interrupted by 'wild intensities' that provided a peek into the ordinarily unrecorded 'quotidian elements of existence'. Michel Foucault, 'Lives of Infamous Men' in *Essential Works of Foucault 1954–1984*, vol. III, ed. by James D. Faubian, trans. by Robert Hurley et al. (New York, NY: New Press, 2001), pp. 157–175, p. 158, p. 170, p. 175. I discuss this essay in more detail in Hannah Proctor, 'History from Within: Identity and Interiority', *Historical Materialism*, 26, 2 (2018), 75–95.

[34] See Svetlana Boym and Marina Mogilner, 'Kak Sdelana "Sovetskaia Sub'etivnost"?' [How is "Soviet Subjectivity" Made?], *Ab Imperio*, 3 (2002), 285–296 and Anna Krylova, 'The Tenacious Liberal Subject in Soviet Studies', *Kritika*, 1, 1 (2000), 119–146. On the tendency of historians to rely on assumptions about psychology without acknowledging the provenance or underpinning of those assumptions, which have histories of their own, see Peter Gay, *Freud for Historians* (Oxford: Oxford University Press, 1985).

Soviet psy-ences discussions of historically situated theories of subjectivity can thus sometimes become confusingly entangled with the author's own (sometimes disavowed) assumptions about human behaviour. Boris Groys has gone so far as to bombastically declare psychology as a discipline fundamentally incompatible with the Soviet system:

> Psychology is religion and the chief occupation of societies, such as those in the West, in which individual private souls are strictly separated from each other through an external legal and economic system of relations. In contrast, the Soviet state intervened directly in the souls of its subjects and manipulated their impressions, feelings and experiences.[35]

Psychology, for Groys, necessarily relies on a conception of autonomous individuated subjectivity at odds with his understanding of Soviet experience in which the 'territory of the psyche that was congruent with the territory of the state. In the Soviet era', he writes, 'private psychology was subordinated to official ideology and therefore was also nationalised'.[36] In his polemic *Homo Sovieticus* (2017) Wladimir Velminski, despite unearthing some fascinatingly eccentric and esoteric strains of Soviet scientific practice, similarly claims that Soviet individuals were effectively hypnotised by the state to produce and reproduce total ideological conformity.[37] In both cases the complex histories of Soviet psy-ences are subordinated to a sweeping characterisation of the supposedly homogenous psychologies of Soviet people. *Psychologies in Revolution* traces one particular strand of the history of the Soviet psy-ences that challenges these reductive portrayals. Psy-ences in the Union of Soviet Socialist Republics (USSR) were indeed subordinated to official ideology (although that by no means resulted in a single monolithic approach and the demands and discourses of the state shifted substantially between 1917 and 1991), but the pages of Luria's books attest to continued incongruities between the psychic territories he attempted to navigate and the maps he employed; incongruities that ultimately, I argue, led him to develop a new mode of scientific writing that

[35] Boris Groys, 'Privatizations/Psychologizations' in *History Becomes Form: Moscow Conceptualism* (Cambridge, MA: MIT Press, 2010), pp. 57–68, p. 58.

[36] Groys, 'Privatizations', p. 59.

[37] Wladimir Velminski, *Homo Sovieticus: Brain Waves, Mind Control, and Telepathic Destiny* (Cambridge, MA: MIT Press, 2017). I discuss this book in more detail in Hannah Proctor, 'Book Review: Homo Sovieticus', *History of the Human Sciences*, September 5, 2017, https://www.histhum.com/413/.

found a different way of describing the diverse psychic terrains he encountered.

Psychologies in Revolution contends that in contrast to the short cases and reports on experimental protocols that characterised many of Luria's works—found in the presentation of his research with criminals conducted in the 1920s through to his later publications such as *Human Brain and Psychological Processes*—he ultimately developed a mode of writing capable of attending to psychic experiences in all their variety, specificity and complexity: the 'romantic' case history. In a review article published in 1973 which prompted Luria to make contact with his American colleague for the first time, Oliver Sacks characterised Luria as a 'divided man' whose work was split between texts 'marked by a certain impersonality and coldness of style which match the "objectivity" of approach they embody' and those which convey 'the infinite complexity of human nature and human beings ... the texture of human thought, perception and action, the ways this can be damaged or disordered, and the ways it can be reconstituted'.[38] In his response Luria rebutted this characterisation insisting that 'only the style of these two books [his two 'romantic' case histories *The Mind of a Mnemonist* and *The Man with a Shattered World*] is different from others, but the principle remains the same'.[39] This may have been an accurate description of his clinical practice, but *Psychologies in Revolution* contends that the sharp difference in written style Sacks observed did mark a decisive shift in the mode in which Luria presented the people with whom he worked as a clinician and thus, at least from a reader's perspective, also marked a shift in principle.

Cracked Monoliths

In the obituary that appeared in *The New York Times* 60 years after the revolution, Luria was described as: 'A Communist Party member who remained aloof from Marxist-Leninist ideology'.[40] This statement implies that Luria maintained a contradictory relationship to the Soviet establishment: paying lip service to the regime, while remaining intellectually

[38] Oliver Sacks, 'The Mind of AR Luria', *The Listener*, June 28, 1973, pp. 870–873, p. 871.

[39] Luria to Sacks, July 19, 1973, Oliver Sacks Archives, Oliver Sacks Foundation, New York City.

[40] 'A.R. Luria, Soviet Psychologist, 75; A Pioneer in Studies of the Brain', *New York Times*, August 17, 1977, p. 12.

autonomous. This narrative, as Loren Graham discusses, was typical of the Cold War era. Graham identifies a tendency in Western writings on Soviet science to emphasise the distinctly Soviet qualities of scientists deemed unsuccessful, dangerously ideological or bogus, while downplaying the social, political and cultural factors that influenced the work of scientists, like Luria, whose work achieved international acclaim.[41] In obituaries that appeared in *The Times*, Luria was similarly defined in opposition to a stereotyped image of the Soviet psychologist as a coldly mechanistic 'rigid Pavlovian'. Instead he was hailed for his concern for 'the acting and suffering individual'; his work was said to exhibit an 'ultimate concern with the human condition', at odds with mainstream Soviet practices.[42] Although Ivan Pavlov was the most influential Soviet psy-entist,[43] whose work on conditioned reflexes was deemed compatible with dialectical materialism and provided the dominant paradigms within the Soviet psy-ences, and Luria was not a Pavlovian (something, as we shall see, he was forced to publicly repent for in the 1950s),[44] Luria's research was not therefore conducted outside of Soviet history. As Joseph Wortis noted in his 1950 overview of the heated debates and heterogeneous methods that characterised Soviet psychology in the wake of the October Revolution:

> The Soviet Union is sometimes depicted as a monolithic giant permanently embedded in fixed Marxian dogma. A reading of Soviet psychiatric literature, however, does not convey the impression of a rigid application of fixed

[41] Loren R. Graham, *What Have We Learned About Science and Technology from the Russian Experience?* (Stanford, CA: Stanford University Press, 1998), pp. 5–8.

[42] See 'Professor A. R. Luria', *The Times*, September 5, 1977, p. 14. See also, Oliver Zangwill, 'Professor A. R. Luria', *The Times*, September 9, 1977, p. 16.

[43] See Daniel P. Todes, *Ivan Pavlov: A Russian Life in Science* (Oxford: Oxford University Press, 2014) and Benjamin Zajicek, 'Scientific Psychiatry in Stalin's Soviet Union: The Politics of Modern Medicine and the Struggle to Define 'Pavlovian' Psychiatry, 1939–1953', PhD thesis, University of Chicago, 2009.

[44] In his first letter to Oliver Sacks Luria took issue with Sacks' claim in a review article on Luria that he studied with Pavlov: 'I never was a pupil of Pavlov, and never studied under him, as a matter of fact I met Pavlov twice in my life—both times for a very short time—no more than half an hour.' Luria to Sacks, July 19, 1973, Oliver Sacks Archives, Oliver Sacks Foundation, New York City. On a trip to the USSR in 1958 (not long after the 'Pavlovian Sessions') the American psychiatrist Henry Murray recalled Luria being scathing about Pavlov's work, saying: 'Pavlov's concepts are not adequate to account for the neurodynamics of man though they are all right for animals.' Henry A. Murray Papers, Harvard University Archives, Miscellaneous Personal Papers and Biographical Notes, 1902–1988, Box 5, Memories of Places visited, USSR, 1958, p. 3.

> formulas. The tempo of change and development, the periods of trial and error, the reversals of policy and the constant atmosphere of experiment and growth are nowhere more apparent than in the fields of Soviet psychology and psychiatry.[45]

Luria's obituary writer was not incorrect to identify a warmth and concern for others in Luria's writings, but those qualities arose within not in spite of their Soviet context. In his analysis of the late Soviet period, *Everything Was Forever, Until It Was No More* (2006), anthropologist Alexei Yurchak challenges accounts of the Soviet experience that carve out a false binary between public utterance and private belief of the sort evident in the obituaries of Luria that appeared in the English-language press. Instead, Yurchak insists that late Soviet life was characterised by paradox: fear, coercion and cynicism coexisted with a commitment to community, creativity and concern for the future.[46] Christina Kiaer argues in relation to the history of art that Soviet culture and society were consistently 'more dynamic and varied than is usually supposed',[47] a statement that could also extend to Soviet science and to Soviet people.

The October Revolution had an impact on individual psychological experience and influenced research agendas in the psy-ences. Luria's work was explicitly addressed to the problem of the mutually transformative relationship between human consciousness and society. He contributed to debates about human development that raged in the

[45] Joseph Wortis, *Soviet Psychiatry* (Baltimore, MD: Williams & Wilkins, 1950), pp. x–xi. See also: Artur V. Petrovsky, *Psychology in the Soviet Union: A Historical Outline*, trans. by Lilia Nakhapetyan (Moscow: Progress, 1990), p. 362. This characterisation is also borne out by more recent scholarship, see, for example: Eugene Raikhel and Dörte Bemme, 'Postsocialism, the psy-ences and mental health', *Transcultural Psychiatry*, 53, 2 (2016), 151–175 and Sarah Marks and Mat Savelli eds., *Psychiatry in Communist Europe* (Houndmills: Palgrave Macmillan, 2015).

[46] See, Alexei Yurchak, *Everything Was Forever, Until It was No More: The Last Soviet Generation* (Princeton, NJ: Princeton University Press, 2006). Jochen Hellbeck similarly challenges such hermeneutic approaches by stating that a 'division between inner striving and outward compliance no longer suffices to understand the self-transformative and self-awakening power of Soviet revolutionary ideology'. Jochen Hellbeck, *Revolution on My Mind: Writing a Diary Under Stalin* (Cambridge, MA: Harvard University Press, 2006), p. 11. Also, for part of this trend in scholarship on Soviet subjectivity, see Igal Halfin, *Terror in My Soul: Communist Autobiographies on Trial* (Cambridge, MA: Harvard University Press, 2003).

[47] Christina Kiaer, 'Was Socialist Realism Forced Labour? The Case of Aleksandr Deineka in the 1930s', *Oxford Art Journal*, 28, 3 (2005), 323–334, p. 323.

aftermath of the October Revolution.[48] Walter Benjamin likened the whole atmosphere of Moscow in 1927 to a laboratory, in which 'no organism, no organisation' was untouched by the experimental fervour.[49] Maria Gough's *The Artist as Producer: Russian Constructivism in Revolution* (2005) quotes this passage in its opening paragraph. For Gough, whose work is focused on avant-garde art practices, experimentation in the early Soviet period was 'not a process sequestered in a laboratory, but rather pervades every aspect of everyday life'.[50] Constructivist artists working in post-revolutionary Russia did not intend to confine themselves to galleries, but they hoped that radically transforming quotidian objects might transform human subjects in turn. In an essay in *LEF* (the journal of the Left Front of the Arts), discussed favourably by Vygotsky in *The Psychology of Art* (1924), Nikolai Chuzhak proclaimed: 'There are no more "temples" of art, or shrines, where the sacred absolutes of priests reside, shrouded in incense. There are workshops, factories, mills, and streets.'[51] Histories of Soviet psychology tend to remain sequestered within laboratories, but Soviet psychologists and psychiatrists were swept up in the revolutionary tumult with everything and everyone else. Indeed, Luria's work was not always confined to laboratories—the studies that form the basis of this book's chapters saw him and his teams conducting research in universities, police cells, rural schools, streets, collective farms, Uzbek tea houses and field hospitals.

Luria's research was bound up with what T.J. Clark has described as the 'buzz of voices, all rattled and contradictory' unleashed by the October

[48] For overviews of Soviet psychology stretching back to the Cold War era, see Raymond Bauer, *The New Man in Soviet Psychology* (Cambridge, MA: Harvard University Press, 1952), David Joravsky, *Russian Psychology: A Critical History* (Oxford: Blackwell, 1989), Alex Kozulin, *Psychology in Utopia: Toward a Social History of Soviet Psychology* (Cambridge, MA: MIT Press, 1984), Matza, *Shock Therapy*, pp. 37–65, John Mcleish, *Soviet Psychology: History, Theory, Content* (London: Methuen, 1975), A.V. Petrovskii, *Psikhologiia v rossii: XX Vek* [*Psychology in Russia: Twentieth Century*] (Moscow: Izd-vo URAO, 2000) and Joseph Wortis, *Soviet Psychiatry* (Baltimore, MD: Williams & Wilkins, 1950).

[49] Walter Benjamin, 'Moscow', *Selected Writings*, vol. 2, ed. by Michael W. Jennings, Howard Eland and Gary Smith, trans. by Rodney Livingstone et al. (Cambridge, MA: Belknap Press of Harvard University Press, 2005), pp. 22–46, p. 28.

[50] Maria Gough's, *The Artist as Producer: Russian Constructivism in Revolution* (Berkeley, CA: University of California Press, 2005), p. 1.

[51] Nikolai Chuzhak, 'Under the Banner of Life-Building (An Attempt to Understand the Art of Today)' trans. by Christina Lodder, *Art in Translation*, 1, 1 (2009), 119–151, p. 143. Originally published in *Lef*, 1 (1923), 12–39.

Revolution, but he also continued to participate in Soviet society long after the 'state started shouting through the revolution's mouth'.[52] His career began in the early years of the Soviet experiment, a moment of optimism and uncertainty, when the question of what a Marxist approach to psychological questions could be was still unsettled and multiple competing theories emerged, but his work continued into the Stalinist period when the parameters of Soviet scientific discourse began to narrow. Although the state sought to contain the contradictory voices buzzing in the wake of the revolution, it would be a mistake to imagine the individuals engaged in producing artistic or scientific work in the Stalin era as nothing more than ventriloquist's dummies. The state had multiple mouths to shout through which could never achieve a fully harmonious or synchronised noise but retained a cacophonous aspect, however, muffled. Indeed, even when Luria himself attempted to produce work within the approved parameters of state-sanctioned discourse, he consistently conducted research with people who failed to do so due to their social experiences or psychological conditions: the Uzbek peasant unfamiliar with the term 'Trade Union', the street child incapable of defining 'The Soviet Union' or the brain-injured Red Army soldier who could no longer recognise Lenin's face. Such people had little in common with Luria's ideal 'advanced' human subject, nor were they the kinds of figures likely to be found striding across Bolshevik propaganda posters or carved in stone atop the pedestals of city squares, but they were Soviet people nonetheless.

It would be possible to cast Luria himself as the victim of an oppressive regime, as a self-interested cynic consciously manipulating the official rhetoric to his own professional advantage or as a genuinely committed Marxist who wilfully overlooked the flaws of the Soviet state. *Psychologies in Revolution* is less interested in Luria's personal motivations than in the texts he produced and the broader historical environment from which they arose, but the contradictions evident in Luria's professional trajectory are worth sketching for what they reveal about the complexities and changing circumstances that influenced how scientific research in the Soviet Union was conducted more broadly. They also point to the limitations of the kinds of reductive and monolithic characterisations of the Soviet experience elaborated by Groys and Welminski.

[52] T.J. Clark, *Farewell to an Idea: Episodes in the History of Modernism* (New Haven, CT: Yale University Press, 1999), p. 297.

In *Stalinist Science* (1997), Nikolai Krementsov discusses the relationship between Soviet scientists and the state, declaring that 'the ***nomenklatura*** system forced practically all Soviet scientists who occupied any administrative post to participate actively in ongoing ideological campaigns in order to maintain their high position'.[53] Yet as was the case with the artists who continued producing work under Stalinism discussed by Kiaer, Luria and his collaborators were 'creators as well as victims'.[54] As we shall see, Luria was not immune from ideological campaigns and he faced public criticism at various points during the Stalin era: he recanted his early interest in psychoanalysis, faced denunciation for being overly reliant on Western 'bourgeois' theories and insufficiently interventionist in the 1930s and lost his position at the Medico-Genetic Institute in 1936. In 1952 he was called upon to publicly repent as part of the notorious 'Pavlovian Session' at the Academy of Agricultural Sciences when all Soviet psychologists were forced to demonstrate their allegiance to Ivan Pavlov's theorisations of conditioned reflexes.[55] He also narrowly avoided persecution in the antisemitic Doctor's Plot in the early 1950s.[56]

[53] Nikolai Krementsov, *Stalinist Science* (Princeton, NJ: Princeton University Press, 1997), p. 258. The infamous case of the Soviet geneticist Trofim Lysenko historically functioned as the paradigmatic example of the detrimental impact of an explicitly ideological approach to nature (see Dominique Lecourt, *Proletarian Science?: The Case of Lysenko* (Atlantic Highlands, NJ: Humanities Press, 1977) and Ethan Pollock, 'From *Partiinost'* to *Nauchnost'* and Not Quite Back Again: Revisiting the Lessons of the Lysenko Affair', *Slavic Review*, 68, 1 (2009), 95–115). Discussions of the pathologisation and imprisonment of dissidents in psychiatric institutions that emerged in the 1960s dominated Western representations of Soviet psychological practices in the Cold War era (see, e.g., Sidney Bloch and Peter Reddaway, *Russia's Political Hospitals: The Abuse of Psychiatry in the Soviet Union* [London: Victor Gollancz, 1977]). This scandal tended to cast the USSR as a uniquely punitive and oppressive state rather than considering parallels with contemporaneous psychiatric practices in the West or considering psychological developments in the Soviet context that did not participate in those repressive programmes. The literature on the political abuse of psychiatry in the Soviet Union is very large, see, for example: A.S. Prokopenko, *Bezumnaia psikhiatriia: Sekretnye materialy o primenenii v SSSR psikhiatrii v karatel'nykh tseliakh* (Moscow: Sovershenno sekretno, 1997), Rebecca Reich, *State of Madness: Psychiatry, Literature and Dissent After Stalin* (DeKalb, IL: Northern Illinois University Press, 2018) and Theresa C. Smith and Thomas A. Oleszczuk, *No Asylum: State Psychiatric Repression in the Former USSR* (New York, NY: New York University Press, 1996).

[54] Kiaer, p. 324.

[55] Levitin, p. 40.

[56] These episodes are succinctly recounted in, Eugenia Kuzoleva and J.P. Das, 'Some Facts from the Biography of A.R. Luria', *Neuropsychology Review*, 9, 1 (1999), 53–56. Levitin claims that during the Doctor's Plot Luria kept a suitcase packed in case he was arrested and

Soon after Stalin's death in 1953, however, Luria began to re-establish his international contacts.[57] In February 1958 he received a letter from the American Zionist writer Horace Kallen who had visited the psychologist in Moscow three decades earlier:

> It is years since I have had a communication so welcome. You have been in our minds over and over again since 1929 and we had been wondering what had happened to you. To see your handwriting at the end of almost thirty years and to learn that you have been filling a Chair in Psychology in Moscow University is to experience something like a friend's or a child's rebirth.[58]

Having fallen out of touch with Luria in the early 1930s, it seems that Kallen, reading news of the Stalinist purges from afar, had feared the worst. Luria would have had reason to fear the worst himself. In July 1936, the decree of the Central Committee of the Communist Party of the Soviet Union (CPSU)(b) 'On Pedological Distortions in the Commissariats of Education' was issued, followed by a series of articles attacking psychologists.[59] In December 1936, Luria was dismissed from his existing institutional positions in Moscow and took on a low-ranking position as a physician. Levitin claims that the 'corridors of the Institute of Neurosurgery saved the assistant Luria during the terrible years of the Terror',[60] suggesting that Luria's position in a medical institution sheltered

reports that his colleague Homskaya resigned her position in Luria's laboratory when concerns were raised by authorities about the number of Jewish doctors employed there, Levitin, 'Epilogue: Luria's Psychological Symphony', *Journal of Russian and East European Psychology*, 36, 6 (1998), 33–62, p. 39. For a review of literature on the 'Doctor's Plot', see: David Brandenberger, 'Stalin's Last Crime? Recent Scholarship on Postwar Soviet Antisemitism and the Doctor's Plot', *Kritika*, 6, 1 (2005), 187–204.

[57] On intellectual exchange across the iron curtain (and between different Eastern bloc countries), see Sarah Marks and Mat Savelli, 'Communist Europe and Transnational Psychiatry' in *Psychiatry in Communist Europe* (Houndmills: Palgrave Macmillan, 2015), pp. 1–26.

[58] Horace Kallen to Luria, February 20, 1958. Cincinnati, OH, American Jewish Archives (AJA), Horace Kallen Papers, MS-1, Box 19, Folder 18.

[59] 'O pedologicheskikh izvrashcheniiakh v sisteme narkomprosov (4 iiulia 1936 goda)' in *KPSS v rezoliutsiiakh i resheniiakh s"ezdov, konferentsii i plenumov TsK*, vol. 6 (Moscow: Politizdat, 1985), pp. 364–367, translated as 'On Pedological Distortions in the Commissariats of Education. A Resolution of the Central Committee of the Communist Party of the Soviet Union July 4, 1936', in Joseph Wortis, *Soviet Psychiatry*, pp. 242–245.

[60] Levitin, p. 43. See also Akhutina, p. 6 and A.A. Leontiev, 'The Life and Creative Path of A.N. Leontiev', *Journal of Russian and East European Psychology*, 43, 3 (2005), 8–69.

him from further persecution. His sister, the wife of a People's Commissar, was arrested in 1937 and sent to a labour camp. Their father, Roman Al'bertovich Luria, who worked as the Kremlin's doctor and treated high-ranking officials involved in the show trials, eventually intervened and she was sent into exile instead.[61]

Focusing solely on these moments in Luria's career (and in Soviet history), however, would be to overlook the extent to which Luria also benefited from the enormous prestige attached to science by the Soviet regime. Indeed, even in the early 1930s, at the very moment his work was being publicly denounced, Luria was offered the opportunity to set up an independent institute in the Ukrainian city of Kharkiv equipped with 16 rooms, funded positions for 15 co-workers and a budget of ₽100,000 per year.[62] In 1947, shortly after the Second World War, he became a full member of the Academy of Pedagogical Sciences of the USSR and in 1951 he was awarded the Order of Lenin, one of the USSR's highest decorations.[63]As he had during the 1920s, from the 'thaw' years on Luria again travelled to international conferences in the West. In the 1960s he was awarded honorary degrees by Leicester University and Nijmegen University in The Netherlands and became an honorary member of the British Psychological Association, Swiss Psychological Society, Columbian Psychological Society and French Neurological Society.[64] Although he relied on his contacts abroad to supply him with products that were scarce in the Soviet Union—including fountain pen cartridges, medicines and camera film—in the post-Stalin period he also had access to luxuries unavailable to his foreign colleagues. Many of his most celebrated late works, for example, including his autobiography and case histories, were completed during periods of relaxation at state-run rest houses situated on grand estates confiscated from the long-since exiled aristocracy.[65]

Regardless of Luria's personal political views, he conducted work within the Soviet system and elaborated a psychological theory broadly consistent with, though not reducible to, Marxist-Leninism. *Psychologies in Revolution*

[61] Levitin, pp. 41–42.

[62] Luria to Wolfgang Köhler, March 6, 1932, Philadelphia, PA, American Philosophical Society (APS), Wolfgang Köhler Papers, Mss.B.K815.

[63] Homskaya, p. 125.

[64] A.R. Luria, 'A.R. Luria' in *A History of Psychology in Autobiography*, vol. 6, ed. by Gardner Lindzey (Cambridge, MA: Harvard University Press, 1974), pp. 253–292, p. 255.

[65] See, for example, Luria to Bruner, December 20, 1969 and Luria to Bruner, May 26, 1976, HUA, HUG 4242.5, Box 88.

contends that Luria is an exemplary Soviet figure precisely because his career and theoretical approach resists neat categorisation according to long-established historiographical paradigms: neither Stalinist lackey nor heroic dissident, neither fervent revolutionary nor defiant anti-Marxist, neither cynical nor sincere (or at least a contradictory and shifting combination of the two).[66] Such binaries are equally insufficient when considering the test subjects and patients Luria and his collaborators interviewed, conducted experiments with and treated.

Paths of Development: Methodology and Structure

In its treatment of Luria's vast oeuvre my approach is intensive rather than extensive: each of the ensuing chapters is structured around close analyses of specific publications. Although the chapters follow the overall arc of Luria's career, this book is not intended as a comprehensive overview of Luria's life and work, says relatively little about the various institutions and disciplines with which he was involved and does not concentrate on his contributions to advancing understandings of the brain for which he is most renowned. Instead, it attempts to situate his publications in their broader historical context by focusing on five key figures around which he chose to structure series of experiments: the criminal, the 'primitive', the child, the aphasic and the synaesthete. These figures were defined according to their cognitive capacities and sometimes by their social positions; their lives were also shaped by specific moments and events in Soviet history. Rather than focusing on archival documents,[67] the emphasis of *Psychologies in Revolution* is on Luria's published material, reflecting my interest in the different forms Luria chose to present his clinical material, in the genre of the case history and in Luria's development of 'romantic' science.[68] The core primary materials that form the basis for the book are

[66] Anna Krylova persuasively claims that since the Cold War era historians' conclusions about Soviet subjectivity might be understood in terms of their divergent interpretations of the relationship between the state and the individual, the official and the unofficial, the public and the private, the sincere and the cynical; despite differences in emphasis (relating to political persuasion etc.) she claims that these binaries remain consistent.

[67] Luria's archives in Moscow—including papers at Moscow State University and his private family archives—remain uncatalogued. Although they have been consulted by various scholars, an in-depth exploration of their contents remains to be published.

[68] Works that have informed my methodological approach to reading case histories and other social scientific or medical documents against the grain include: Saidiya Hartman,

not untranslated or unpublished materials, revealing some hitherto overlooked episode in the history of the Soviet psy-ences. Instead, my analysis deliberately focuses on some of Luria's most well-known publications, long available in English translation, as I hope to demonstrate that attending to the manifest content of these works allows for an analysis of their latent contexts.[69]

The key texts that form the basis of my first three chapters are: *The Nature of Human Conflicts* (published in English in 1932), *Cognitive Development* (based on research conducted in Central Asia in 1931 and 1932, but unpublished in either Russian or English until the 1970s) and a cluster of texts on children and childhood based on research conducted between 1923 and 1936. These works follow a similar format: they include descriptions of experiments undertaken, transcriptions of responses to protocols and general theoretical analyses and conclusions. They also each focus on specific psychological 'types' that Luria defined in opposition to 'advanced' human consciousness: convicted murderers, Uzbek people with little or no formal education and children respectively. Reading these works I was struck that the short cases and responses Luria cites, which provide glimpses into the everyday lives of Soviet people, often seemed to exceed or complicate his accompanying interpretations of the material. In these publications, despite constantly asserting his interest in tracing the ways human consciousness is shaped by history in theory, he does not always give a detailed demonstration of how this happens in practice.

The final chapters of the book focus on Luria's two well-known case histories: *The Man with a Shattered World* (1971) and *The Mind of a Mnemonist* (1968). *Psychologies in Revolution* argues that it was only in these case histories that Luria developed a form of scientific writing capable of fully attending to the utterances and experiences of the people he dedicated his career to observing, understanding and treating. Unlike the majority of his publications, these works are written in a detailed and empathetic style alert to the particularities of the conditions of the people

Wayward Lives, Beautiful Experiments: Intimate Histories of Social Upheaval (New York, NY: W.W. Norton, 2019), Rebecca Lemov, *Database of Dreams: The Lost Quest to Catalog Humanity* (New Haven, CT: Yale University Press, 2015) and Edith Sheffer, *Asperger's Children: The Origins of Autism in Nazi Germany* (New York, NY: W.W. Norton, 2018).

[69] Much of the recent secondary literature on Vygotsky consists of careful philological reconstruction aimed at restoring the integrity of his original works and rectifying distortions that arose when his work was edited and translated. In this book I have opted to take Luria's existing publications at face value as I am less interested in intentions than in effects.

they describe rather than seeking to relate those people's consciousnesses to some abstract normative ideal. These publications do not represent a shift in scientific practice so much as a shift in the stylistic presentation of his clinical material.[70] Luria characterised this as a distinction between 'classical' and 'romantic' science. By taking a 'romantic' approach to Luria's more 'classical' writings, I hope to discern moments in which snatches of 'the wealth of living reality' gleam through the cracks of his more detached prose.[71] The book's final sections contend that the attentiveness to history achieved in Luria's literary or 'romantic' case histories constitutes one of his major theoretical contributions and that his fine-grained analysis of experiential specificity might be the aspect of his approach most capable of challenging the very different (though no less ideologically shaped) psychological assumptions, paradigms and practices of the present, a question that is explicitly addressed in the book's conclusion.

Psychologies in Revolution begins with the 1917 October Revolution, which Luria also framed as the beginning of his own biographical and professional journey. *The Nature of Human Conflicts*, which was first published in English translation in 1932, contains a wealth of clinical material based on Luria's experiments designed to assess responses to emotional stimuli conducted in the 1920s, including a series of tests undertaken with recently convicted murderers. Unlike his sometimes bombastically written theoretical assessments of psychoanalysis written during the same period, *The Nature of Human Conflicts* provides insights into the concrete (and frequently distressing) experiences of Soviet people and thus complicates the optimistic arguments regarding social transformation he made elsewhere. **The Criminal** argues, through a close analysis of published case histories, that Luria documented psychic conflicts, antagonisms and tensions in practice even when he sought to overcome them in his more abstract, theoretical writings.

In 1931 and 1932 Luria led two expeditions to Soviet Central Asia. Conducted during the period of the First Five Year Plan (1928–1932), the experiments were conceived as an attempt to trace the cognitive effects of that transformative state policy. Luria defined Uzbek society as culturally

[70] Luria outlines the form of 'syndrome analysis', fundamental to neuropsychology, that also lay at the foundations of 'the psychology of personality' explored in his romantic case histories in 'A.R. Luria', *History of Psychology in Autobiography*, p. 285.

[71] Luria, *Making of Mind*, p. 174.

backward, but he hoped to describe a process of psychic advancement, which he assumed would accompany the collectivisation of agriculture, mass literacy campaigns, campaigns against religion and rapid industrialisation taking place at that moment in the region. As with his critique of psychoanalysis, this research affirmed social and individual progress even though his research findings told a more complicated story. **The 'Primitive'** confronts Luria's engagement with the Stalinist project and discerns moments of tension his publications document between his research team and the Uzbek people they attempted to categorise, classify and comprehend. It concludes by focusing on the figure of the *ichkari* woman—Muslim women with no formal education who wore the traditional veil (*paranja*)—who Luria defined as the most cognitively 'backward' and 'passive' group among his test subjects. Transcriptions of responses to his experiments contained in *Cognitive Development* are read as sites of contestation in which the people being interviewed seem to exceed Luria's assumptions about them.

As an icon of the future, the child was central to Soviet ideology. Untainted by the bourgeois past yet still dependent on adults, the question of how to create the conditions for a properly revolutionary childhood was a paradox faced by Soviet psychologists and educators alike. **The Child** focuses on Luria's experiments with children between 1923 and 1936. It considers his theoretical discussions of early infancy, discusses a series of experiments conducted with twins which probed the relationship between nature and nurture and finally considers a text focused on children from distinct social backgrounds. Whilst working at the Academy of Communist Upbringing, Luria undertook comparative studies of children from three different social groups. 'Speech and Intellect among Rural, Urban and Homeless Children' (1930) includes a comparative study of the urban city school child, the rural child and the 'urchin' child. The latter group, like the '*ichkari*' women discussed in the previous chapter, was defined as the least 'advanced' of the three groups and was the slowest to respond to psychologists' questions. Abandoned children (*besprizornye* or *besprizorniki*) were a prominent feature of Soviet urban life. The bourgeois nuclear family was being challenged for ideological reasons, but many children were separated from their parents by historical circumstances. Although homeless children existed in Russia prior to the October Revolution, in the wake of the violence, famine and epidemics that accompanied the First World War and the Civil War, their numbers greatly increased after it. This chapter discusses how Luria's understanding of the ideal development of the child was complicated by the actual children he encountered who were growing up in situations far removed from his laboratories.

The Second World War, or 'Great Patriotic War' as it became known in the Soviet Union, marked a decisive turning point in Luria's career. Much of his post-war research was conducted with people who had endured serious brain injuries on the battlefield. In 1971 Luria published a case history of a patient he had been observing for nearly 30 years. *The Man with a Shattered World* tells the story of Zasetsky, a Red Army soldier in the Second World War who survived a bullet wound to his head. Unlike the works analysed in the book's previous chapters in which details of individual cases are confined to one of two paragraphs, *The Man with a Shattered World* was composed in a self-consciously literary style in a warmth-infused tone that emphasises the rapport between patient and doctor. Drawing on literary methodologies, **The Aphasic** considers the formal qualities of Luria's case history and proposes that the mode of writing Luria developed allowed him to present the cognitive condition of his wounded patient in all its complexity without framing healing as a means of overcoming 'backwardness'. Through this mode of 'romantic' scientific writing *Psychologies in Revolution* argues, Luria was able to foreground his patients rather than his schema for understanding them. In his discussion of case histories, John Forrester notes that 'the promise of the entirely revealed life, in its singularity and distinctiveness, is undoubtedly still alive in the devotees of [psycho]analysis' and that 'the genre of the case history…sustains that desire',[72] but he also observes a tension between this pursuit of particularity and the function of the medical case as an exemplar, a tension between the specific and the general, the concrete and the abstract or (in Luria's terms) the romantic and the classical that pulses through Luria's publications and also animates my discussions of them.

The book ends by returning to the immediate post-revolutionary period with which it began. **The Synaesthete** considers the friendship Luria maintained with the Soviet filmmaker Sergei Eisenstein from the mid-1920s until the director's death in 1948. Both Eisenstein and Luria lived and worked through the post-revolutionary period and despite experiencing criticism and censorship, survived the Stalinist purges (1937–1938) and continued to work through the Second World War (Eisenstein

[72] John Forrester, 'If p, then what? Thinking in cases', *History of the Human Sciences*, 9, 3 (1996), 1–25, p. 10.

died in 1948). Their careers existed in fraught relationship to the Soviet state and its erratic yet prescriptive vocabularies and policies; both of their projects fail to dovetail neatly with dominant narratives for understanding the production of cultural and scientific work in the Soviet Union. Drawing on materials from the Eisenstein archives, this concluding section follows the paths of their contemporaneous careers and considers their correspondence with one another, discussing moments of convergence and divergence between Luria's and Eisenstein's intellectual projects, in relation to the developmental movements and progressive models for understanding human consciousness around which *Psychologies in Revolution* is structured. Existing analyses of Eisenstein and Luria's relationship tend to explore the influence of the psychologist on the artist, whereas my discussion insists that the inspiration flowed in two directions and examines how the formal discussions between the two men informed Luria's 'romantic' approach to scientific writing with a particular focus on his case history *The Mind of a Mnemonist*, the protagonist of which Eisenstein also developed a relationship. *Psychologies in Revolution* then concludes by proposing that Luria's 'romantic' science offers a valuable model for scientific writing that could still be drawn on in the present situating his work in relation to contemporary critiques of neuroscientific discourse.

CHAPTER 2

The Criminal

In his reflections on Alexander Luria's life and work, first published shortly after Luria's death, Karl Levitin recalled asking the neuropsychologist: 'Whom does the neuropsychologist most closely resemble—the physician, the research scientist, the experimenter, the armchair theoretician, or all of them together?' In response to this question Luria, who had long been a lover of detective novels, replied: 'A criminal investigator'.[1] By piecing together disparate clues to solve a case, the work of the detective functioned as an apt metaphor for Luria's scientific research, but it wasn't only a metaphor. In his early career in Moscow in the 1920s, Luria briefly pursued his own criminal investigations with people accused of murder, in collaboration with prosecutors in Moscow, in an attempt to ascertain if 'the psychologist studying the affective traces in the criminal [could] objectively establish his participation in the crime'.[2]

This chapter focuses on Luria's *Nature of Human Conflicts: Emotion, Conflict and Will* (1932),[3] based on experiments conducted between

[1] K.E. Levitin, 'The Soul's Frail Dwelling-House', *Journal of Russian and East European Psychology*, 36, 5 (1998), 3–45, p. 27. Originally published in Russian as *Mimoletnyi uzor* [*A Transient Pattern*] (Moscow: Izdatel'stvo Zhanie, 1978).

[2] A.R. Luria, *The Nature of Human Conflicts or Emotion, Conflict and Will*, trans. by W. Horsley Gantt (New York, NY: Liveright, 1932), p. 78.

[3] The book was initially set to be published under the English title 'Human Nature and its Conflicts', to which Luria objected given the book's lack of focus on human nature. This resulted in the shuffled word order. He had proposed as alternatives: '"Affection [sic], Conflict & Will" or "Psychophysiology of affection [sic] and conflicts"'. Luria to W.H. Gantt,

H. Proctor, *Psychologies in Revolution*, Mental Health in Historical Perspective, https://doi.org/10.1007/978-3-030-35028-4_2

1923 and 1930, which he described in a letter to his friend Sergei Eisenstein in 1930 as a work on 'criminals, examinations, internal neuroses'.[4] The book was published in a slightly clunky English translation with no Russian version appearing in Luria's lifetime.[5] He recalled that in his second and last meeting with Pavlov in Leningrad in 1932, he brought a copy of the English book with him which was received disdainfully. He humorously recounted the exchange in a letter written many years later to Oliver Sacks:

> "I shall not read the book, it is too big … and you are a psychologist?" "Yes, sir" "That's bad: psychologists are so stupid! Everybody knows that one has to start from simple facts and then move to complex, and psychologists can even not understand this!"[6]

Luria's book only devotes one chapter to the experiments with criminals, and also encompasses discussions of experiments with university students undergoing examinations, children,[7] people with aphasia and Parkinson's, as well as a series of experiments based on hypnotic sessions with people

April 10, 1931, Alan Mason Chesney Archives, Johns Hopkins University, WH Gantt Papers, B45 F28.

[4] Luria to Sergei Eisenstein, February 17, 1930, Moscow, Russian State Archives of Literature and Art (RGALI), Sergei Eisenstein Papers, 123-1-19321/26.

[5] In 1931 Luria wrote the following oblique message to his English translator: 'You ask about the Russian edition of my book. It, unfortunately, has not yet happened; I hope that it will appear in the winter of this year.' Luria to W.H. Gantt, August 17, 1931, Alan Mason Chesney Archives, Johns Hopkins University, WH Gantt Papers, B45 F28. Reflecting on his research with criminals in the 1970s he claimed that a huge unpublished manuscript remained from the experiments, which he said he sent to the Institute for Criminal Investigation at that time. He mentions his boss K.N. Kornilov's increasing hostility to psychoanalysis as the reason he was forbidden from publishing another article based on some of these experiments, which may also have been a factor in the fate of the publication of the book, Levitin, pp. 71–72. The publication of *The Nature of Human Conflicts* in English was preceded by the publication of articles in German: A.R. Luria, 'Die Methode der abbildenden Motorik bei Kommunikation der Systeme und ihre Anwendung auf die Affektpsychologie', *Psychologische Forschung*, 12 (1929), 127–179 and A.R. Luria, 'Die Methode der abbildenden Motorik in der Tatbestandsdiagnostik', *Zeitschrift für angewandte Psychologie*, 35 (1930), 139–183.

[6] Luria to Sacks, July 19, 1973, Oliver Sacks Archives, Oliver Sacks Foundation, New York City.

[7] These experiments were conducted by Mark Lebedinsky, who would go on to work with Luria and others in Kharkiv. See Anton Yasnitsky and Michel Ferrari, 'From Vygotsky to Vygotskian psychology: Introduction to the history of the Kharkov School', *Journal of the History of the Behavioral Sciences*, 44, 2 (2008), 119–145. Luria also collaborated with Lebedinsky on the experiments with people with aphasia and Parkinson's at the laboratory of the Clinic for Nervous Diseases at the First Moscow University.

diagnosed with hysteria and other neurotic conditions.[8] The diverse experiments that form the basis of *The Nature of Human Conflicts* were conducted during the same period with which Luria was involved with the Russian Psychoanalytic Society in Moscow. Though he also drew on a range of other influences, this chapter considers the short case histories and accounts of experiments conducted with murderers included in *The Nature of Human Conflicts* in relation to Luria's simultaneous engagement with psychoanalysis, asking to what extent they were informed by his readings of psychoanalytic theories and interpretations of psychoanalytic methodologies.

In 1924 Luria delivered a speech to the Russian Psychoanalytic Society. The resulting paper 'Psychoanalysis as a System of Monistic Psychology', in which Luria argued for the compatibility of psychoanalysis with Marxism, was included in K.N. Kornilov's 1925 collection *Psychology and Marxism*. In this essay Luria emphasised the importance of acknowledging the impact of social environments on individuals, which he expressed in Marxist terms: 'the human mind is a product of the brain and, in the last analysis the effect of the social environment and the class relations and conditions of production underlying it.'[9] Luria argued that an approach to psychology grounded in historical materialism ought to take both biological and sociological factors into consideration, treating consciousness as historically and socially determined, rather than as something essential and eternal. Psychoanalysis, he argued, was the only school of psychology that met these criteria because it was devoted to studying 'the whole individual, his behaviour, inner workings, and motive forces ... instead of "extra-social" man in general, studied in abstraction from the social conditions forming him'.[10] Placing such theoretical proclamations alongside publications based on Luria's clinical work from the same period brings tensions between his theories and practices into focus, conflicts that are further explored through a comparison of Luria's methods and conclusions with

[8] For a rough chronology of the experiments and papers the book comprises and shifts in influence and terminology evident across the text, see Michael Paul George Hames, 'The Early Theoretical Development of Alexander Luria', PhD, University of London, UCL, 2002, pp. 149–156.

[9] A.R. Luria, 'Psychoanalysis as a System of Monistic Psychology', *Journal of Russian and East European Psychology*, 40, 1 (2002), 26–53, p. 29. Originally published in K.N. Kornilov, *Psikhologiia i Marksizma* [*Psychology and Marxism*] (Leningrad: Gosudarstvenoe psikhologii, 1925), pp. 47–80.

[10] Luria, 'Psychoanalysis...', p. 14.

those of other Soviet psychological studies that emerged from this period of cooperation between criminology and psychiatry. But before arriving in Moscow during the period of the New Economic Policy (NEP) (1921–1928), it is worth briefly summarising Luria's early career in Kazan in the immediate aftermath of the October Revolution.

Luria's Early Career: Kazan 1917–1923

In 1918, a year after the October Revolution, Luria entered Kazan University, which, seven years later, would be renamed after its most famous alumnus: V.I. Lenin. Luria claims that his family, who were part of the city's Jewish intelligentsia, were 'sympathetic to the revolutionary movement [though] not directly involved in it'.[11] Indirect involvement with the revolution was unavoidable, however, as Luria reflected: 'The Revolution freed us, especially the younger generation, to discuss new ideas, new philosophies and new social systems. … The content and the style of our lives changed almost immediately.'[12] The revolution interrupted routine existence; in its immediate aftermath food was scarce, money worthless, railway connections disrupted and industry stagnant. The year 1918 was a tumultuous year in Luria's native city, a crucial strategic site in the ongoing Civil War in which Bolshevik victory was still far from assured. During the summer, the White Army occupied Kazan, forcing the Bolsheviks into retreat. Any Reds remaining were shot on sight. On 10 September, the Red Army stormed the city and the Whites fled across the Volga.[13] Leon Trotsky, who led the attack on Kazan, depicted victory as a major turning point in the Civil War. The day after the Red Army's triumph, he delivered a rousing speech in the city's main theatre, declaring: 'We value science, culture, art, and want to make them accessible to the people, along with all their institutions—schools, universities, theatres and the rest. … Precisely for that it was that we fought under the

[11] Luria, *The Making of Mind: A Personal Account of Soviet Psychology*, ed. by Michael Cole and Sheila Cole (Cambridge, MA: Harvard University Press, 1979), p. 18.

[12] Luria, *Making of Mind*, pp. 18–19.

[13] On Kazan's strategic importance in the Civil War, see Evan Mawdsley, *The Russian Civil War* (New York, NY: Pegasus Books, 2005), p. 67 and W. Bruce Lincoln, *Red Victory: A History of the Russian Civil War* (New York, NY: Simon and Schuster, 1989), pp. 187–193.

walls of Kazan.'[14] Even in the midst of war, educational and cultural pursuits were explicitly situated at the heart of the revolutionary project.

Luria depicted his early academic career as being shaped by the October Revolution, which had already begun to have a concrete impact on the cultural institutions mentioned by Trotsky.[15] At Luria's university there were new admission policies, shifts in curricula and changes of staff. Quotas that restricted the number of Jewish students permitted to attend university were abolished. Courses continued intermittently if not uninterruptedly. Professors unable to adapt to the demands of the new regime often allowed students to lead discussions. Luria admits to having 'no idea of the real causes of the Revolution' nor any knowledge of Marxist theory at this time. Instead he describes the revolution as a vividly experienced yet nebulous 'atmosphere' of 'general excitement'[16]—chaotic, hopeful and experimental. Kristin Ross's characterisation of Marx's understanding of the 1871 Paris Commune might be fruitfully applied here: the significance of the revolution was felt 'in its displacement of the political onto seemingly peripheral areas of everyday life—the organisation of space and time, changes in lived rhythms and social ambiances'.[17] Luria recalled that

> my friends and I immediately threw our whole beings into the new movement because we recognised the opportunities that it offered. My enthusiasm came more from a strong emotional, romantic feeling toward the events of the time than from any deep intellectual appreciation of their social roots. ... Instead of cautiously groping for a foothold in life, we were suddenly faced with many opportunities for action—action that went far beyond our tiny circle of family and friends. The limits of our restricted, private world

[14] Leon Trotsky, 'The Significance of the Taking of Kazan in the Course of the Civil War', *The Military Writings and Speeches of Leon Trotsky, Vol 1, 1918: How the Revolution Armed*, trans. by Brian Pearce (London: New Park Publications, 1979), p. 332.

[15] On the revolution's impact on academic institutions, see Michael David-Fox, *Revolution of Mind: Higher Learning Among the Bolsheviks, 1918–1929* (Ithaca, NY: Cornell University Press, 1997). Loren Graham characterises the post-revolutionary period in scientific research institutions as being a 'curious amalgam of political control and popular ferment'. Loren Graham, *Science in Russia and the Soviet Union* (Cambridge: Cambridge University Press, 1993), p. 93.

[16] Luria, *Making of Mind*, p. 19.

[17] Kristin Ross, *The Emergence of Social Space: Rimbaud and the Paris Commune* (London: Verso, 2008), p. 33.

> were broken down by the Revolution, and new vistas opened before us. We were swept up in a great historical movement.[18]

Luria was drawn to utopian socialism and to discussions of 'politics and the shape of the future society'.[19] In this lively and confused period of intellectual eclecticism, Luria notes that he 'developed a lasting aversion to psychology'.[20] He found nothing 'alive' in the dry, academic works of Wilhelm Wundt, Hermann Ebbinghaus, Edward B. Titchener and other 'classical' Western European psychologists whose work had a neo-Kantian underpinning: 'the boredom they inspired ... is quite beyond fathoming'.[21] He felt that the forms of laboratory-based studies dominant in mainstream psychology were abstract, removed from everyday life and thus incapable of capturing the rich complexity of 'people as they live and behave in the world'.[22] It was at this moment that he first encountered the work of Sigmund Freud, in whose writing he found an antidote to the psychological works that had dominated the university syllabus; Freud's work seemed equal to the vitality of the revolutionary moment. In 1922 Luria, then only 19 years old, organised a psychoanalytic discussion group in Kazan, writing to Freud personally to inform him that 'a new organisation had appeared in the world, the Kazan Psychoanalytic Circle'. Freud responded, addressing the young Russian enthusiast as 'Herr President', and authorising Luria to translate his work into Russian.[23]

The Kazan Psychoanalytic Society was thus founded, with Luria acting as secretary.[24] Luria's report in the Bulletin of the International Psycho-Analytic Association from 1923 lists 22 members, mostly medical practitioners and students, who, he declares, 'interest themselves in the theoretical problems of psychoanalysis and in its application to the mental and social sciences as well as to therapeutics'.[25] The group met at least

[18] Luria, *Making of Mind*, p. 19.

[19] Luria, *Making of Mind*, p. 20.

[20] Karl Levitin, 'A Criminal Investigation', *Journal of Russian and East European Psychology*, 36, 5 (1998), 46–75, p. 63.

[21] Levitin, 'A Criminal Investigation', p. 63.

[22] Luria, *Making of Mind*, p. 22.

[23] Levitin, 'A Criminal Investigation', pp. 62–65.

[24] For an account of the activities of the Kazan Psychoanalytic Society drawing on papers from the Luria family archive, see van der Veer and Valsiner, *Understanding Vygotsky: A Quest for Synthesis* (Oxford: Blackwell, 1991), pp. 78–82.

[25] Luria, 'The Kazan Psycho-Analytical Society, Russia', *Bulletin of the International Psycho-Analytic Association*, 4 (1923), 397–399, p. 397.

twice a month in its first year. Discussions included analyses of specific cases, dream interpretations, the relationship of psychoanalysis to other psychological schools and summaries of specific works by Freud (including 'On Narcissism' and 'Group Psychology and the Analysis of the Ego'). Luria's talks primarily functioned to situate psychoanalysis within the broader discipline of psychology in Russia, including discussions of reflexology and neurology. He presented psychoanalysis as one approach to psychology among many devoted to 'the analysis of personality (as a whole)'.[26] The Kazan Psychoanalytic Society declared an interest in combining psychoanalytic techniques with other approaches and was far from uncritical of the psychoanalytic method. At a meeting on 23 November 1922, for example, doubts were raised as to the applicability of psychoanalytic categories to literature and folklore. Another debate focused on childhood sexuality and the question was raised in discussing Freud's 'Group Psychology and the Analysis of the Ego' (first published in 1921) whether sexuality was an appropriate paradigm for considering collective dynamics, a pressing question in post-revolutionary Russia.[27]

Luria was enrolled in medical school and simultaneously took classes at the Pedagogical Institute at Kazan Psychiatric Hospital. As there was no psychological laboratory in Kazan at that time, he accepted a position in a laboratory at the Kazan Institute for the Scientific Organisation of Labour where he completed studies into fatigue with foundry workers. The poet and trade unionist Aleksei Gastev had established a Communist brand of Taylorism at the Central Institute of Labour in Moscow, founded in 1920, advocating for the Scientific Organisation of Labour [*Nauchnaia Organizatsiia Truda/NOT*], devising methods to maximise productivity and efficiency, which informed the work being undertaken in Kazan.[28] Luria's experiments with workers in Kazan are an early example of the ways his research was shaped and informed by specifically Soviet discourses and practices, including those with which he was not

[26] Luria, 'Kazan', p. 397.

[27] Luria, 'Kazan', p. 399.

[28] On Soviet Taylorism, Aleksei Gastev and industrial psychology, see Rainer Traub 'Lenin and Taylor: The Fate of "Scientific Management" in the (Early) Soviet Union', *Telos*, 37 (1978), 82–92, Kendall E. Bailes, 'Alexei Gastev and the Soviet Controversy over Taylorism, 1918–1924', *Soviet Studies*, 9 (1977), 373–94 and R.S. Schultz and R.A. McFarland, 'Industrial psychology in the Soviet Union', *Journal of Applied Psychology*, 19, 3 (1935), 265–308.

directly associated or those of which he may not have been an explicit proponent.

Alongside his engagement with psychoanalysis, Luria also became familiar with the work of the leading figures in Russian psychiatry and neurology: Ivan Pavlov and Vladimir Bekhterev. Luria travelled to Petrograd to ask the latter to join the editorial board of a journal he and some colleagues had co-founded, who agreed on condition that they slightly alter its title. Paper was a scarce resource in post-revolutionary Kazan and Luria resorted to printing some of the copies of this journal on the discarded yellow wrappings from a local soap factory.[29] These 'soap-paper articles'[30] attracted the attention of K.N. Kornilov who invited Luria to Moscow to join the Moscow Institute of Psychology of which Kornilov had recently become the director. Luria moved to Moscow in 1923 to take up this position where he also became Scientific Secretary of the Russian Psychoanalytic Society based in a recently constructed mosaic and stained glass festooned art nouveau mansion confiscated from the industrialist Stefan Riabushinsky (later assigned to Maxim Gorky, whose museum the building still houses today).[31] An Institute for Psychoanalysis had formed in Moscow in 1922, making the Soviet capital the first place outside of Vienna and Berlin where it was possible to train as a psychoanalyst.[32]

[29] Levitin, 'A Criminal Investigation', p. 67.

[30] Luria, *Making of Mind*, p. 28. A slightly different account is given in a 1974 speech transcribed by Levitin in which Luria claims to have written to Kornilov to inform him of the experiments he was conducting on reaction time. Levitin, 'A Criminal Investigation', p. 67.

[31] On the 'International Solidarity' psychoanalytic children's home run by Vera Schmidt in the building, see Vera Schmidt, *Psychoanalytische Erziehung in Sowjetrussland: Bericht über das Kinderheim-Laboratorium in Moskau* [*Psychoanalytic Education in Soviet Russia: Report on the Children's Home-Laboratory in Moscow*] (Leipzig: Internationaler Psychoanalytischer Verlag, 1924) and Yorkdanka Valkanova, 'The Psychoanalytic Kindergarten Project in Soviet Russia, 1921–1930', *SCRSS Digest*, 13, 2 (2016), 12–15. On the building, see William Craft Brumfield, *The Origins of Modernism in Russian Architecture* (Berkeley, CA: University of California Press, 1991), pp. 133–139, N.L. Penezhko, V.N. Chernukhina and A.M. Marchenkov, *6 Riabushinsky Gorku—Dom Na Maloi Nikitkoi, 6* [*6 Riabushinsky Gorky—The House at 6 Mal. Nikitskaia*] (Moscow: Nasledie, 1997), p. 74 and James L. West, *Riabushinsky's Utopian Capitalism' in Merchant Moscow: Images of Russia's Vanished Bourgeoisie* (Princeton, NJ: Princeton University Press, 2014), p. 165.

[32] Martin A. Miller, *Freud and the Bolsheviks: Psychoanalysis in Imperial Russia and the Soviet Union* (New Haven, CT: Yale University Press, 1998), p. 59.

Objective Psychoanalysis

Luria acted as Secretary of the Russian Psychoanalytic Society in Moscow from 1923 to 1927, remaining a member until the society's activities came to a halt at the end of the decade.[33] In 1932, Luria published a penitent article in which he distanced himself from his early interest in Freud. He characterised psychoanalysis in terms consistent with the prevailing Soviet critique: for being individualised, pessimistic and irrational. He attacked Freud for straying from his clinical experiences and criticised psychoanalysis for its emphasis on 'the primitive, prehistorical biological roots of the human personality'.[34] Mark Solms has argued that although the hostility to Freudian thought propounded by the Stalinist regime forced Luria to renounce his early affiliations with psychoanalysis, his work continued to be psychoanalytically oriented: 'the apparent change in Luria's scientific direction initially amounted to little more than a change in *terminology*.'[35] René van der Veer and Jaan Valsiner similarly indicate that Luria retained an interest in psychoanalysis throughout his career despite the official blacklisting of the subject in the Soviet Union, an observation that chimes with recent historiography challenging the assumption that psychoanalytic concepts and practices completely disappeared from the psy-ences in the Soviet sphere after 1930.[36]

[33] 'The psychoanalytic movement slowed down, and about the year 1930 came to a standstill. From this date it officially ceased to exist, and all publication of its work ceased likewise', Elias Perepel, 'The Psychoanalytic Movement in U.S.S.R.', *Psychoanalytic Review*, 26 (1939), 298–300, p. 300. For a representative selection of essays from the period both defending and attacking psychoanalysis, see *Zigmund Freid: Psikhoanaliz i russkaia mysl'* [*Sigmund Freud: Psychoanalysis and Russian Thought*], ed. by V.M. Leibin (Moscow: Izdatel'stvo Respublika', 1994), pp. 145–372. On the fate of psychoanalysis in Soviet Russia, see Alberto Angelini, 'History of the Unconscious in Soviet Russia: From its Origins to the Fall of the Soviet Union', *International Journal of Psychoanalysis*, 89 (2008), 369–388, V.N. Buzin, 'Psychoanalysis in the Soviet Union: On the History of a Defeat', *Russian Social Science Review*, 36, 6 (1995), 65–73, Alexander Etkind, *Eros of the Impossible: The History of Psychoanalysis in Russia* (Boulder, CO: Westview Press, 1997), Gary N. Goldsmith, 'Between Certainty and Uncertainty: Observations on Psychoanalysis in Russia', *Journal of Analytical Psychology*, 47 (2002), 203–224 and Miller, pp. 53–113.

[34] A. Lurii a, 'Krisis burzhuaznoi psikhologii' [The Crisis of Bourgeois Psychology], *Psikhologiia*, 1–2 (1932), 63–88, p. 66.

[35] Mark Solms, 'Freud, Luria and the Clinical Method', *Psychoanalysis and History*, 2 (2000), 76–109, p. 92.

[36] René van der Veer and Jaan Valsiner, *Understanding Vygotsky: A Quest for Synthesis* (Oxford: Blackwell, 1991), p. 87. See, Sarah Marks, 'Suggestion, Persuasion and Work:

Yet although Luria's article was certainly motivated by the oppressive political climate and continuities between his early and later work are evident, it would be equally misleading to claim that his psychological approach was ever primarily psychoanalytic.[37] His engagement with psychoanalysis, even at its most publicly laudatory, always combined psychoanalysis with other psychological approaches and read it through a Marxist lens. For Luria in the mid-1920s, psychoanalysis' emphasis on the interrelation between mind, body and the ever-changing environment distinguished it from other approaches to the mind. Luria argued that Marxism and psychoanalysis both looked 'beyond the surface of things to their real roots'.[38] Delving into the unconscious implied a practical and materialist approach to psychic life, firmly grounded in a belief in the primacy of history (though the introduction to the Russian translation of Freud's *Beyond the Pleasure Principle* Luria co-wrote with Vygotsky in 1925, by focusing a text in which Freud insists on instinctual conservatism, also indicated the limits of the compatibilities between Marxism and psychoanalysis Luria attempted to identify).[39] As with the group discussions in Kazan, the topics of Luria's lectures at the Russian Psychoanalytic Society in Moscow indicate that he tended to incorporate psychoanalytic approaches with

Psychotherapies in Communist Europe', *European Journal of Psychotherapy and Counselling*, 20, 1 (2018), 10–24, pp. 13–15, which particularly emphasises the continued presence of psychoanalytic practices in parts of the Eastern bloc that came under Communist control in the aftermath of the Second World War. On the use of psychoanalytic therapeutic methods in Soviet psychiatry during the Second World War, see Benjamin Zajicek, Scientific Psychiatry in Stalin's Soviet Union: The Politics of Modern Medicine and the Struggle to Define 'Pavlovian' Psychiatry, 1939–1953, PhD thesis, University of Chicago, 2009, p. 148. Luria wrote the entry on psychoanalysis for the 1940 *Great Soviet Encyclopedia*, which is descriptive and measured rather than vehement in its criticisms. A. Luriia, 'Psikhoanaliz', *Bolshaiia Sovietskaia Entsiklopediia*, tom. 47 (Moscow and Leningrad, 1940), pp. 186–187.

[37] Given Luria's own remarks about the profound effect meeting Vygotsky in 1924 had on his intellectual development, it is also worth noting how closely Luria's criticisms of psychoanalysis in his repentant article recall those articulated by Vygotsky, who argued that psychoanalysis 'directly reduces the higher mental processes—both personal and collective ones—to primitive, primordial, essentially prehistorical, prehuman root, leaving no room for history'. Lev Vygotsky, 'The Historical Meaning of The Crisis in Psychology' in *The Collected Works of Lev Vygotsky*, vol. 3, ed. by Robert W. Rieber and Jeffrey Wollock, trans. by René van der Veer (New York, NY: Plenum Press, 1997), pp. 233–343, p. 263.

[38] Luria, 'Psychoanalysis…', p. 47.

[39] For a detailed discussion of this introduction, see my article Hannah Proctor, 'A Country Beyond the Pleasure Principle: Alexander Luria, Death Drive and Dialectic in the Early Soviet Union', *Psychoanalysis and History*, 18, 2 (2016), 155–182.

discussions of conditioned reflexes and motor reactions. On 26 March 1925, for instance, Luria delivered a paper on 'Affect as an Unabreacted Reaction', which combined a discussion of Freud's theories of sexual inhibition with observations on Pavlov's experiments with dogs.[40] *The Nature of Human Conflicts* is an example of Luria's attempts in this period to create a new form of 'objective psychoanalysis',[41] combining methods emerging from Soviet psychology with insights from psychoanalysis and German Gestalt psychology in an effort to record affective experience empirically, drawing less on Sigmund Freud than on Carl Jung.

Jung does not feature heavily in existing secondary accounts of psychoanalysis in early twentieth-century Russia, primarily appearing through his connection with the Russian psychoanalyst Sabina Spielrein, his patient and lover in Zurich who subsequently trained as a psychoanalyst and whose paper 'Destruction as the Cause of Coming into Being' (first published in German in 1912) inspired Freud's concept of the death drive.[42] Though there were plans to translate Jung's works into Russian, it seems none were published before the closure of the state Institute of Psychoanalysis.[43] Spielrein returned to the Soviet Union from Western Europe in 1923 where she joined the Russian Psychoanalytic Society, the same year Luria became its secretary. Alexander Etkind speculates that the young provincial Luria would have been in thrall to Spielrein, but there is little evidence about the content of their discussions in Moscow and it does not seem she was responsible for introducing him to Jung's work.[44]

[40] Luria, 'Russian Psychoanalytic Society', *Bulletin of the International Psycho-Analytic Association*, 7 (1926), 294–295, p. 294.

[41] Levitin, 'A Criminal Investigation', p. 68.

[42] On Spielrein see, John Launer, *Sex versus Survival: The Story of Sabina Spielrein, Her Life, Her Ideas, Her Genius* (Lexington, KY: Lulu, 2011) and Angela Sells, *Sabina Spielrein: The Woman and the Myth* (New York, NY: SUNY Press, 2017). In the years preceding the October Revolution, Russian psychiatrists were influenced by and paid visits to Jung. See, Irina Sirotkina, *Diagnosing Literary Genius: A Cultural History of Psychiatry in Russia, 1880–1930* (Baltimore, MD: Johns Hopkins University Press, 2002), p. 103, pp. 142–143.

[43] Miller claims that the Institute for Psychoanalysis intended to translate and publish Freud's 'most influential works', followed by works by Jung, Sandor Ferenczi and Melanie Klein, *Freud and the Bolsheviks*, p. 60. Etkind discusses Spielrein's involvement in translating Jung's works into Russian, claiming that she was initially wary as she hadn't read Russian translations of psychoanalytic works so was unsure of the terminology, *Eros of the Impossible*, p. 165.

[44] Etkind, p. 173.

Luria claims in his autobiography that he discovered Jung and Adler around the time he first read Freud's *The Interpretation of Dreams* (1900), while still living in Kazan. Specifically he mentions being 'deeply impressed' by Jung's *Studies in Diagnostic Associations* (1906) and states that he saw Jung as offering a more materialist approach to psychology than Freud:

> Many of Freud's ideas seemed speculative and somewhat fantastic to me, but the study of emotional conflicts and complexes using the method of associations seemed promising. Here, I thought, was a scientific approach that combined a strongly deterministic explanation of concrete, individual behavior with an explanation of the origins of complex human needs in terms of natural science.[45]

Luria's experiments with workers conducted at the Institute for the Scientific Organisation of Labour in Kazan, which measured reaction times to verbal stimuli, already bore a Jungian influence.[46] It was these kinds of methods—including word association tests, recordings of muscle tension and measurements of reaction times—that Luria continued to develop upon his arrival in Moscow and that formed the basis of *The Nature of Human Conflicts.*

Luria arrived at the Moscow Institute of Psychology in the aftermath of the October Revolution when the question of what a properly Marxist approach to the discipline should be was still unsettled and contested. In old age he reflected on the youth of the staff: few at the institute were older than 24. Enthusiasm thus took the place of experience. Kornilov was a proponent of 'reactology' and the entire institute was structured around this new discipline.[47] Reactology was intended as a materialist counterpart to the bourgeois idealism that had characterised pre-revolutionary approaches to psychology.[48] In a 1928 overview of 'Psychology in Russia',

[45] Luria, *Making of Mind*, p. 23.

[46] Luria, *Making of Mind*, p. 25. For a description of these experiments, see also Levitin, 'A Criminal Investigation', p. 67.

[47] Kornilov may have participated in denunciations of psychoanalysis but soon after reactology was similarly 'proclaimed an erroneous deviation from truly Marxist psychology and denounced in June 1931, by the decision of the Institute's local cell of the Communist Party', Anton Yasnitsky, *Vygotsky: An Intellectual Biography* (London: Routledge, 2018), epub, p. 328.2.

[48] On Luria's work with Kornilov at the Moscow Institute of Psychology and its relation to his psychoanalytic work, see Alex Kozulin, *Psychology in Utopia: Toward a Social History of Soviet Psychology* (Cambridge, MA: MIT Press, 1984), pp. 86–89 and Levy Rahmani, *Soviet*

Luria described Kornilov as having 'established the basic principle of what he calls the unipolar expenditure of energy',[49] developing an understanding of 'the influence of the social conditions of the subject' in an attempt to apply 'the doctrine of dialectical materialism to psychology'.[50] Decades later, however, Luria dismissed Kornilov's approach as 'naive, naturalistic, and mechanistic', based on the erroneous assumption 'that there was a fixed amount of energy in an organism that had to be shared by the mental and motor systems; the more energy expended on the mental component of an action, the less remained for its motor component.'[51] Kornilov's definition of energy as a finite resource overlapped with concurrently elaborated Soviet theories of sexuality which, in a transformation of the Freudian concept, redefined sublimation as physiological rather than psychical, proposing that committed revolutionaries should rechannel the energy expended performing activities such as non-reproductive sex, which was seen as 'wasteful', individualistic and frivolous, into productive work or revolutionary organising on behalf of the collective.[52]

Kornilov's methods involved measuring motor reactions using special devices and drawing assumptions about 'mental energy' from reaction times or the strength and duration of reactions. Despite Luria's scepticism about Kornilov's definition of energy, he also devised experiments that involved measuring reactions using various devices, including a kymograph, which measures blood pressure. Though he had worked with a dynamoscope in Kazan, a kind of stethoscope for measuring muscle tension, in Moscow he came to view it as 'too crude an instrument' and replaced it with an 'Ermakov apparatus', 'to study the dynamic components

Psychology: Philosophical, Theoretical and Experimental Issues (New York, NY: International Universities Press, 1973), pp. 25–30.

[49] Luria, 'Psychology in Russia', *Pedagogical Seminary and Journal of Genetic Psychology*, 35 (1928), 347–355, p. 349.

[50] Luria, 'Psychology in Russia', p. 350.

[51] Luria, *Making of Mind*, p. 25.

[52] On the understanding of sublimation prevalent in Soviet psychology and in NEP-era discourse more broadly, see Frances Lee Bernstein, *The Dictatorship of Sex: Lifestyle Advice for the Soviet Masses* (DeKalb, IL: Northern Illinois University Press, 2011), pp. 151–156 and Hannah Proctor, 'Reason Displaces All Love', *New Inquiry*, February 2014, in which I discuss relationships between Taylorist understandings of workplace efficiency and the sexual theories of psychologist Aron Zalkind, including *Revoliutsiia i molodezh'* [*Revolution and Youth*] (Moscow: Izdanie Kommunistich. un-ta im. Sverdlova, 1925) and *Polovoi vopros v usloviiakh sovetskoi obshchestvennosti* [*The Sexual Question in Soviet Social Conditions*] (Leningrad: Gosizdat, 1926).

of writing', deeming this a 'much more "alive"' device than those favoured by Kornilov.[53] Combining these physiological methods with insights from psychoanalysis, Luria proceeded to develop an approach that used the analysis of motor reactions to draw conclusions about 'inner, emotional conflicts'.[54] Such an approach would not separate the physical from the psychological but instead proposed that a connection existed between the two realms. His experiments were based on the assumption that analysing the body could give insights into the mind and even into the unconscious, giving a material grounding to what he saw as the more speculative aspects of psychoanalytic theory.

Although Luria wrote to the English translator of *The Nature of Human Conflicts* in 1931 requesting that references to psychoanalysis and behaviourism be removed from the blurb on the cover of the book and any publicity materials, disavowing the connection between his publication and those psychological approaches (presumably in response to the suspicion with which they had come to be viewed in the Soviet Union),[55] he was later explicit in acknowledging the influence of Jung on his research during that period:

> Free associations as used by Jung in his *Studies of Diagnostic Associations* (1910) was one component of the technique we developed. We asked the subject to engage in a motor response simultaneously with each verbal associative response. … We had in mind the kind of phenomena Freud and the psychoanalytic school were interested in, emotion-laden experiences that motivate and guide people's behavior far beyond the boundaries of the experience itself. We began by devising a laboratory model of the problem as we thought it would occur in real life situations.[56]

Jung's tests are based on the assumption that word associations 'are not meaningless and that these are not random choices out of the thousands of possible reactions, but just those that indicate the ideas occupying the foreground of the individual interest'.[57] Similarly, Luria assumed that the

[53] Levitin, p. 68.

[54] Luria, *Making of Mind*, p. 32.

[55] Luria to W.H. Gantt, April 10, 1931, Alan Mason Chesney Archives, Johns Hopkins University, WH Gantt Papers, B45 F28.

[56] Luria, *Making of Mind*, p. 32.

[57] Carl Jung, *The Collected Works of CG Jung, vol. 2, Experimental Researches*, ed. by Sir Herbert Read, Michael Fordham, Gerhard Adler and William McGuire, p. 321.

responses of people in distressing or unusual circumstances to particular words would be emotionally, and hence physiologically, disturbed or 'disorganised'. He proposed that 'the structure of internal changes' would be 'accessible to direct observation'[58] and that 'affect is reflected in the disorganisation of active behaviour':[59]

> One of the first things we discovered in this work is that strong emotions prevent a subject from forming stable, automatic motor and speech responses, although subjects of equivalent intelligence, operating under normal circumstances, can form such responses after only a few trials. It appeared as if subjects influenced by strong emotions adapted to each new situation in a unique way and did not settle into a stable reaction pattern. Not only did the subjects have unstable motor and speech responses when considered separately, but they seemed unable to create a single functional system that included both motor and speech components, often delaying the speech components of their reactions.[60]

In order to test this hypothesis and to observe these kinds of disturbances, Luria and his team, which included the young Aleksei Leontiev, sought to conduct experiments with people in 'a marked affective state or people whose affective complexes are already activated', including people in situations it was assumed would provoke unusually strong emotional responses.[61]

Following an introductory section introducing the main concepts and methods employed in the book, its second chapter consists of results from a set of experiments conducted with students waiting to undergo examination as part of the Communist Party's 1924 purge (literally cleansing [*chistka*]) of students in the higher schools. This was intended to eliminate counter-revolutionary 'bourgeois' forces from educational institutions, resulting in the expulsion of 18,000 students from Russian institutions.[62] Students were interrogated on their class and academic backgrounds and their political convictions before the commission decided whether to expel

[58] Luria, *Nature*, p. 22.

[59] Luria, *Nature*, p. 43.

[60] Luria, *Making of Mind*, p. 35.

[61] Luria, *Nature*, p. 39. Luria also published a text on these experiments in Russian: *Eksamen i psikhika* [*Exams and the Mind*] (Moscow: Tr. Akademii Kommunisticheskogo Vospitaniia imni Krupskoi, 1931). See also Levitin, p. 70.

[62] See Sheila Fitzpatrick, *Education and Social Mobility in the Soviet Union, 1921–1934* (Cambridge: Cambridge University Press, 1979), pp. 97–105.

them from the institution, an experience Luria assumed had a 'traumatic character'.[63] A second series of experiments were conducted in 1927 with students undergoing ordinary university examinations, which came to similar conclusions, though Luria conceded that the situation provoked less extreme reactions as its stakes were far lower. In the 1924 experiments, Luria and his team plucked people directly from the examination queues in the hope of gaining an insight into the immediate psychic impact of the ordeal. His results focus on the physical agitation and excitability of the students. He presented people with a series of words and found that those most closely pertaining to the anticipated interrogation—including 'examination', 'to surrender' and 'commission'—evoked the most extreme physical responses or motor reactions.[64] Concurring with findings of Max Wertheimer, Luria also found that a failure to respond or a failure to hear the stimuli were common reactions, which he took to indicate a 'disorganised' emotional state.

Luria defined the examination as a traumatic site in which it was possible to observe how 'affect ruptures the organised course of associative processes'.[65] The excitability of people in this situation was contrasted to responses of a control group. The examination was shown not only to have a 'disorganising' impact on people's thought processes but to reduce them to a state 'ordinarily foreign to the cultured adult and usually seen only in the earlier stages of development'.[66] According to Luria, the emotional states produced in people by this situation had a regressive aspect, 'bringing about that state of a person long since past'[67]; making adults regress to childhood or, expressed in phylogenetic rather than ontogenetic terms, to slip back from a 'civilised' to a more 'primitive'[68] psychological level. He went on to observe that the intense affects observed in such situations are similar to neuroses. Indeed, he claimed they temporarily produce a kind of neurosis in the respondent which can only be understood in relation to the individual's general mental state—'normal, healthy'[69] people

[63] Luria, *Nature*, p. 48.

[64] Luria, *The Nature of Human Conflicts or Emotion, Conflict and Will*, trans. by W Horsley Gantt (New York, NY: Liveright, 1932), p. 57.

[65] Luria, *Nature*, p. 51.

[66] Luria, *Nature*, p. 53.

[67] Luria, *Nature*, p. 51.

[68] Luria, *Nature*, p. 52.

[69] Luria, *Nature*, p. 75.

will react differently from people who have a pre-existing 'neurotic disposition'.[70]

In the introduction to the Russian translation of Freud's *Beyond the Pleasure Principle* Luria co-wrote with Lev Vygotsky in 1925, they argued bombastically that a radical transformation in external social conditions would result in a fundamental transformation of human psychology:

> The only forces which make it possible to ... propel us toward progress and activity, are external forces, in our terms, the external conditions of the material environment in which the individual exists. It is they that represent the true basis of progress, it is they that create the real personality and make us adapt and work out new forms of psychic life.[71]

The authors hailed the dialectical qualities of Freud's text but questioned how the 'stormy progression of the historical process' could be explained, if there exists, as Freud insisted, an innate biological compulsion to return to an inanimate state, an instinctual conservatism inherent to every organism.[72] Luria and Vygotsky recognised that Freud's theory 'breaks with teleological concepts', but the alternative model they proposed, informed by an understanding of revolutionary change which would see a radical social transformation transform psychic life in turn, paradoxically failed to retain the elements of Freud's theory they praised for being dialectical in the first place (the ongoing tension between the life and death instincts).[73] Their utopian optimism in this theoretical text is further called into question by Luria's observations in *The Nature of Human Conflicts* that a state policy intended to root out the vestiges of the bourgeois past had a regressive impact on the minds of those people subjected to it, dragging them back to a more 'primitive' level. Rapid social transformation here in practice was said by Luria to occasion a movement into the past rather than propelling society, and hence individual people's individual psychologies, forwards; regression took the place of progression.

[70] Luria, *Nature*, p. 76.

[71] Alexander Luria and Lev Vygotsky, 'Introduction to the Russian translation of *Beyond the Pleasure Principle*' in *The Vygotsky Reader*, ed. by René van der Veer and Jaan Valsiner (Oxford: Basil Blackwell, 1994), pp. 10–18, p. 15. First published in Russian in Sigmund Freud, *Po tu storonu principa udovol'stviia* [*Beyond the Pleasure Principle*] (Moscow: Sovremennye Problemy, 1925), pp. 3–16.

[72] Luria and Vygotsky, 'Introduction…', p. 13.

[73] Luria and Vygotsky, 'Introduction…', p. 15.

Murder Cases

The published notes of the Russian Psychoanalytic Society sometimes hint at the psychic impact of the harsh present on the people Soviet psychoanalysts encountered as their patients. On 13 May 1926, for instance, a paper was delivered in Moscow discussing the case of a patient who had burnt herself to death: 'The writer thought that in this case—that of a girl whom an obsessional impulse drove to commit suicide by burning—illustrates the isolated operation of the destructive instinct.'[74] The speaker's insistence that this was an 'isolated' incident implies that the operation of a destructive instinct was interpreted not as a universal phenomenon or as one that might have a social aetiology.

The third chapter of *The Nature of Human Conflicts* is devoted to a discussion of a set of experiments Luria conducted with convicted murderers. Yet despite his emphasis on the determining impact of social environments on human psychology in many texts from the same period, very little analysis of how that impact might manifest itself in people's thoughts, feelings or behaviour appears in these pages. In 1923 Luria set up a 'special laboratory in Moscow'[75] with the procurator's office where he worked for around five years conducting experiments with approximately 50 people, some of whom were brought to Moscow from Kyiv and almost all of whom were accused of murder. The experiments included in the chapter were conducted between 1925 and 1927.[76] Most contemporaneous Soviet criminologists pursuing psychological examinations of criminals seem to have worked with people already sentenced and imprisoned, whereas Luria and his team experimented with people immediately after their capture by the police 'and before they were told of the cause of their arrest',[77] though their conclusions did not tend to be used in the judicial process of accusation and persecution. Luria recalls being warned that all arrested people would be emotional regardless of whether they were guilty, but he assumed only those with 'memory traces relevant to the crime' would react to words connected to the incident.[78]

[74] Luria and Vera Schmidt, 'Russian Psycho-Analytical Society', *Bulletin of the International Psycho-Analytic Association*, 8 (1927), 454–455, p. 454.

[75] Luria, *Nature*, p. 79.

[76] Hames, p. 151.

[77] Luria, *Nature*, p. 79.

[78] Levitin, p. 71.

As in the purge commission experiments, Luria proceeded by recording reactions to a series of words with a kymograph, concluding that reactions to words which directly referred to the circumstances of the murder provoked the most pronounced responses. The 'combined motor method' he devised went beyond psychoanalysis, which he claimed was 'insufficiently reliable, resting on non-essential and unstable indicators of deep processes occurring below consciousness' by adding an 'objective' physiological component 'uniting inner (verbal-associative) processes, which can be used to merge affective states into a single functional system, with their outer, motor manifestations, to create a single dynamic contour in which the inner changes are necessarily reflected in the outwardly observable motor processes'.[79] He outlined this method of supplementing word association with measurements of physical responses in the presentation he gave at the Ninth International Congress of Psychology at Yale University in 1929:

> The method proposed is intended to be a study of the motor reactions which are connected with associative processes. The subject is told to press with a finger on a pneumatic appliance while giving a verbal response. The disturbance of the motor curve express a corresponding affective change evoked by verbal reactions. ... This method suggests the possibility of recording the psychophysiological structure and typology of all the processes connected with the integration of human behaviour and particularly with the revealing of affects and their traces.[80]

For example, in a case of a murdered woman who was found half-naked in a bin the words 'garbage' and 'dress' agitated the respondent.[81] His experiments with criminals were primarily designed to facilitate confession: 'Attempts to conceal the affective traces accentuate the objective records;

[79] Luria, 'A.R. Luria' in *A History of Psychology in Autobiography*, vol. 6, ed. by Gardner Lindzey (Cambridge, MA: Harvard University Press, 1974), pp. 253–292, p. 258.

[80] Luria, 'The New Method of Expressive Motor Reactions in Studying Affective Traces', *Proceedings and Papers of the Ninth International Congress of Psychology (held at Yale University, New Haven, Connecticut, September 1st–7th 1929)* (Princeton, NJ: Psychological Review Company, 1930), pp. 294–296. Only a summary of Luria's paper is contained here. On Luria's trip to Yale, see Anton Yasnitsky, 'Rekonstruktsiia poezdki A.R. Lurii na IX Mezhdunarodnyi Psikhologicheskii Kongress' [A reconstruction of A.R. Luria's journey to the Ninth International Psychological Congress], *Voprosy Psikhologii* (2012), 86–93.

[81] Luria, *Nature*, p. 116.

confession of the crime (or the catharsis) brings about first the disappearance of the diffused disturbances.'[82] And as with the students awaiting examination, Luria declared the psychic state of the criminal in the aftermath of a murder, particularly in unpremeditated cases, as a regression to an earlier stage of development.[83]

Jung also suggested that his word association tests could be used 'to diagnose by the associations the complex of a crime'.[84] In *Studies in Word Association* (1918) he gives an example in which word association tests were used to establish guilt, following a theft using a method similar to that employed by Luria and his team in Moscow. The suspect was presented with a series of 'critical-stimulus words' that pertained to the situation of the crime mixed up with 'irrelevant or random words'.[85] Convinced that the former group of words demonstrably agitated the subject, Jung proceeded to make an accusation:

> The total result of this experiment appeared so convincing that I told the subject point blank that he was a thief. The young man who, up to now, had shown an embarrassed smiling face, turned suddenly pale and protested his innocence with great excitement. I then pointed out to him several points in the experiment that seemed to me particularly convincing. Thereupon he suddenly burst into tears and confessed.[86]

Jung's criminal is a young man suspected of stealing money from a drawer of shirts and ties. The stimulus words included 'writing case', 'chest of drawers' and 'linen cupboard'.[87] 'New Aspects of Criminal Psychology' (1908) begins with another case of theft in which Jung imagines the following objects being stolen from an Italian hotel:

> a bag containing jewellery, such as a gold bracelet with blue stones, a diamond brooch in the shape of a butterfly, a ring shaped like a snake, and a brooch in the form of a lizard with emerald eyes ... a green leather wallet

[82] Luria, 'New Method...', p. 295.

[83] He likens the criminal to an 'idiot' and a child. Luria, *Nature...*, p. 83.

[84] C.G. Jung, *Studies in Word Association: Experiments in the Diagnosis of Psychopathological Conditions Carried Out at the Psychiatric Clinic of the University of Switzerland*, trans. by M.D. Eder (London: William Heinemann, 1918), p. 299.

[85] Jung, *Studies in Word Association*, p. 334.

[86] Jung, *Studies in Word Association*, p. 341.

[87] Jung, *Studies in Word Association*, p. 334.

> containing a cheque drawn on the Banca Commerciale Italiana and three banknotes of fifty lire each, and a bottle of Odol.[88]

The stimulus words for establishing guilt are plucked from this scenario and include gold, fifty, three, bracelet, blue, bank, snake, stone, diamond, lizard, green, leather, butterfly, wallet, cheque, banknote, Odol and so on. Although the cases described here are far less extreme than those discussed by Luria, Jung's methods in these two examples are very close to Luria's. The middle-class domestic space and opulent environment of an Italian hotel described in these pieces are incidental to Jung's conclusions, just as the shabbiness of Moscow communal apartments are to Luria's. Yet the glittering luxurious objects that appear in Jung's cases stand in such stark contrast to the situations described by Luria that the comparison gestures towards the kinds of social and cultural factors for which the word association method fails to account.

Although Luria mentions details from other murder cases at various points in *The Nature of Human Conflicts*, just five short case histories are included in the chapter explicitly discussing affect in criminals. They all follow a similar format, containing a brief factual description of the crime scene sometimes followed by further details of the incident revealed during questioning. The case of the woman found dead in a bin is relayed as follows:

> On 28th January 1927, in a garbage can in the courtyard of one of the Moscow houses was found the body of a half-dressed woman, whose head had been crushed by some kind of heavy instrument. Investigation traced her back to a dwelling occupied by seasonal workers, where there was living at that time only one person, a house painter. Fil., 47 years old. In the room were found traces of blood, and in the stove an unusual bulk of ashes and the charred remains of a woman's dress. It seemed that the body was carried from the room in the garbage can between 5 and 7 o'clock on the morning of that day. The rigidity of the body, however, made us think that the murder was done at least a day before.[89]

Luria then discusses the subsequent confession in which 'Fil.' revealed that he had invited a woman home with him, drunk a bottle of vodka with

[88] Jung, 'New Criminal Psychology' in *The Collected Works of C.G. Jung, vol 2, Experimental Researches*, ed. by Sir Herbert Read, Michael Fordham, Gerhard Adler and William McGuire (Princeton, NJ: Princeton University Press, 1981), pp. 586–596, pp. 587–588.

[89] Luria, *Nature*, p. 115 (modified translation).

her and fallen asleep only to be woken up by the sounds of her rifling through his pockets. Suspecting her of theft he drunkenly struck her with his hands and then with a large stick, fell back to sleep and awoke again to realise she was dead. In other cases we hear of a baker accused of killing his wife; a man found in a pile of snow having been hit with a sledgehammer; a factory worker who broke a window at his workplace to steal a ventilator; a man who killed his fiancée and threw her dead body into water tied to a cast-iron wheel; and a woman who reported returning home to find her husband in bed with a cut throat and crushed skull but who was eventually accused of arranging for a neighbour to commit the murder in exchange for ₽18 and the husband's coat and cap. The most bizarre and gruesome case Luria describes (discussed in my Introduction) dates from 1927 when a corpse in a basket was delivered to the main railway station in Moscow. After its discovery the husband of the murdered woman received an anonymous note containing a baggage coupon informing him where 'he might obtain his wife'.[90] In this instance, as no one had confessed to the crime, Luria interviewed four suspects and the results of his experiments were used to make a conviction.

In these experiments Luria has no interest in probing why people may have committed murder but only in whether they committed a murder. In the 'combined motor method' procedure employed by Luria and his team, in the absence of a verbal confession, the criminal's physical responses, hesitations or associations give the game away. He had no interest in what may have motivated these peculiar gory acts, nor in their long-term psychic impact on the individuals accused of committing them. Luria's model for understanding human behaviour in these experiments was designed to observe and record agitation and distress but not to interpret or ameliorate those feelings or to probe their origins. In his discussion of acts of sexual violence relayed by Soviet medical professionals from the same period, Dan Healey notes that though cases tended to be reported in a matter-of-fact way, '[v]iscious assaults accompanied by cruelty and indiscriminate violence could move doctors to depart from the dispassionate and laconic prose they usually cultivated in the expert report'.[91] No such departures or deviations from the expert tone are evident in the case histories Luria included in *The Nature of Human Conflicts.* Similarly

[90] Luria, *Nature*, p. 109.

[91] Dan Healey, *Bolshevik Sexual Forensics: Diagnosing Disorder in the Clinic and Courtroom, 1917–1939* (DeKalb, IL: Northern Illinois Press, 2009), p. 98.

abbreviated accounts of individual's experiences appear in the chapters on neuroses, which though he indicated may have had social origins, he failed to analyse explicitly in those terms: 'Subject K., a hysterical woman 30 years old, had been under the observation of psychiatrists [for] some weeks. Her complex was connected with the fact that her husband was an alcoholic, and wasted all his money on drink, and furthermore grossly mistreated her.'[92] Yet although Luria relayed experiences in a detached register, it is still possible to discern in his sparse descriptions of murders, thefts and violent incidents hints of the social conditions of NEP-era Moscow that formed their backdrop: material hardship, insecure working conditions, mistrust between neighbours, workplace insubordination, inebriation, alcoholism and sexual aggression.

Luria's case histories are like tiny iceberg tips poking out of the giant ocean of history. The lack of contextualisation provided in the cases is especially surprising given Luria's own frequent emphasis on the impact of social and environmental factors on psychic life in his theoretical texts from the same period. In 'Principle Questions of Contemporary Psychology', published in the Bolshevik journal *Under the Banner of Marxism* in 1926, Luria declared that

> The developing requirements of practical life have pushed the task of deep self-contemplation, the study of one's own consciousness, etc., far into the background; instead, the task is to understand the mechanism of human actions—their motives, the forces that organise a person's responses to the demands posed by their environment—in short, to understand the mechanics of human adaptation to the environment.[93]

In 'Psychoanalysis as a System of Monistic Psychology' (1925) he framed his arguments in even more explicitly politicised terms:

> For Marxism, the human mind is a product of the activity of the brain, and, in the final analysis, of the effects of the social environment and the class relations and conditions of production underlying it on the brain and on each individual human being. Any attempt to isolate "thought" or "mind"

[92] Luria, *Nature*, p. 423.

[93] Luriia, 'Printsipial'nye voprosy sovremennoi psikhologii' [Principle Questions of Contemporary Psychology], *Pod znamenem marksizma* [*Under the Banner of Marxism*], 4–5 (1926), 129–139, p. 129.

> as some discrete class of phenomena Marxism regards as an unscientific and patently idealistic approach to things.[94]

Yet *The Nature of Human Conflicts* provided little analysis of the class backgrounds, external conditions or social environments of the people Luria discussed who were accused of committing murder.

This apparent contradiction between theory and practice can be explained by looking more closely at what Luria intended when he spoke of history. In the concluding passages of *The Nature of Human Conflicts* Luria declares that

> The behaviour of the human adult is primarily the product of complex growth, which cannot be comprehended as an accumulation of experiences. Human psychology differs from the zoological point of view in that it views specific laws absent in the phenomena of nature and characteristic of history. The development of the human as a historical subject occurs as the elaboration of special forms of historical, cultural behaviour.[95]

History recurs as a theme across his writings but despite his occasional invocations of class—a concept which appears with far less frequency than the more general 'environment' [*sreda*]—when Luria invoked history the term tended to indicate development understood in the longue durée. He was less interested in the psychic impact of the kinds of granular transformations in everyday life [*byt*] that could be observed in Moscow between, say, 1917 and 1927 than he was in understanding what distinguished humans from animals or 'civilised' people from 'primitives', which in turn connected to his interest in how children develop into adults. He was interested in the historical emergence of language, signs and tools (and, correspondingly, in the infant's introduction to these things): 'the permutation of activity by the use of cultural means make the human a new biological series in history'.[96] But he was more interested in language and tools as such than in analysing the psychological impact of, for example, how Russian people's vocabularies changed in the aftermath of the

[94] A.R. Luria, 'Psychoanalysis as a System of Monistic Psychology', *Journal of Russian and East European Psychology*, 40, 1 (2002), 26–53, p. 27. Originally published in K.N. Kornilov, *Psikhologiia i Marksizma* [*Psychology and Marxism*] (Leningrad: Gosudarstvenoe psikhologii, 1925), pp. 47–80.

[95] Luria, *Nature*, p. 428.

[96] Luria, *Nature*, p. 428.

October Revolution or in material shifts associated with life in the NEP period (such as housing conditions, forms of scarcity or the introduction of new kinds of objects) or, indeed, in transformations in the judicial system with which he was cooperating.[97] It is instructive in this regard to compare Luria's methods and the presentation of his clinical material with publications written by contemporaneous Soviet psychologists and criminologists.

Other Voices, Other Rooms

Speaking in 1974 Luria claimed, 'it was very easy' to forge a connection with the police in Moscow in the 1920s: 'we didn't have to waste many years of effort and submit endless petitions to different authorities. I simply went to the Moscow Prosecutor's Office, presented him with our idea, and soon a laboratory was set up.'[98] Luria's recollection is corroborated by Daniel Beer's historical account, which emphasises the 'intellectual and institutional freedom' with which criminologists and psychiatrists pursued their work 'in the period between the Civil War and the Cultural Revolution' during which time many experts from the human sciences offered 'biomedical explanations of crime, social disorder, and subversion'.[99] Sharon A. Kowalsky designates the period between 1922 and 1928 the 'golden age' of Soviet criminology, observing that crime studies participated in evaluating 'the population's progress toward socialism and the distance remaining to the successful construction of a socialist society'.[100] Although Luria's experiments seem unusual in their access to criminals immediately following arrest, Luria was far from the only psychiatrist

[97] For example, there is no sense of how Luria's experiments connected to social order policing policies or to the new criminal code introduced in 1922. It is also unclear whether the criminals Luria came into contact with have been classified by the state as 'socially harmful elements' [*sotsial'no-vredye elementy*]. See, David R. Shearer, *Repression and Social Order in the Soviet Union, 1924–1953* (New Haven, CT: Yale University Press, 2009).

[98] Levitin, p. 70.

[99] Daniel Beer, *Renovating Russia: The Human Sciences and the Fate of Liberal Modernity, 1880–1930* (Ithaca, NY: Cornell University Press), p. 167, p. 168.

[100] Sharon A. Kowalsky, *Deviant Women: Female Crime and Criminology in Revolutionary Russia, 1880–1930* (DeKalb, IL: Northern Illinois University Press, 2009), p. 50, p. 8. See also, Louise Shelley, 'Soviet Criminology after the Revolution', *Journal of Criminal Law and Criminology*, 70, 3 (1979), 391–396 and Peter H. Solomon 'Criminalization and Decriminalization in Soviet Criminal Policy, 1917–1941', *Law & Society Review*, 16, 1 (1981–1982), 9–44.

proximate to the police and judicial system in Russia during the NEP era; expert psychiatric opinions were sought by police in various kinds of cases and various expert analyses were conducted with criminals in Soviet prisons by people working in the psy-ences. In his discussion of rape cases and other crimes involving sexual violence, for example, Healey describes the central role psychiatrists came to play in the Soviet justice system: 'Procedural codes required police and prosecutors to seek psychiatric expert opinions when the defendant in such cases appeared to be mentally ill'[101] and psychiatrists were also called upon to give expert advice on the possible motives for crimes or on imputability.[102]

Although Luria's experiments can be situated within these broader trends, his methods and conclusions seem to have differed from other analyses conducted and published in the same period. Kowalsky opens her book on Soviet criminology with the dramatic case of a 24-year-old woman named Nastia E. who on '1 October 1923 at seven in the morning … maimed her husband by amputating his penis'.[103] The case, which recounted the events leading up to the violent act, which saw the culprit encounter her husband's child with another woman, is discussed by the psychologist A.E. Petrova in an article included in the 1924 collection *The Criminal World of Moscow*, edited by the criminologist M.N. Gernet who was based at the Moscow Bureau for the Study of Criminal Personality and Crime.[104] The same case was also discussed by the psychologist and criminologist Nikolai Brukhanskii in his 1927 work *Materials of Sexual Psychopathology*.[105]

[101] Dan Healey, *Bolshevik Sexual Forensics: Diagnosing Disorder in the Clinic and Courtroom, 1917–1939* (DeKalb, IL: Northern Illinois Press, 2009), p. 11. He describes how 'psychiatrists like all expert doctors now had the right to operate as "judges of scientific facts", to participate in criminal investigations and in court hearings as expert witnesses with nominal independence from police interference', p. 110.

[102] Healey, pp. 113–114.

[103] Petrovna, 'Sluchai izuvecheniia muzha' in *Prestupnyi mir Moskvy*, ed. by M.N. Gernet (Moscow: Pravo i zhizn, 1924), pp. 82–101, p. 82, cited in Kowalsky, p. 3.

[104] On Gernet and the Moscow Bureau for the Study of the Criminal Personality and Crime [Moskovskii kabinet po izucheniiu lichnosti prestupnika i prestupnosti], see Kenneth Pinnow, *Lost to the Collective*, pp. 57–59, pp. 144–145 and 'From All Sides: Interdisciplinary Knowledge, Scientific Collaboration, and the Soviet Criminological Laboratories of the 1920s', *Slavic Review*, 76, 1 (2017), 122–146.

[105] See N. Brukhanskii's *Materialy po seksual'noi psikhopatologii* [*Materials of Sexual Psychopathology*] (Moscow: MM and C Sabashnikovy, 1927), pp. 9–18.

Petrovna's chapter reproduces the case from a 1923 conference and is followed by the author's in-depth psychological analysis of the incident. More than double the length of the cases that appear in *The Nature of Human Conflicts*, the case of Nastia E. contains far more background information about the defendant—where she was born, her parents' professions, her education, her current class position and occupation, her medical history (including accounts of menstruation and venereal disease), her reasons for moving to Moscow, her relationship with her husband and so on. It is then followed by a detailed discussion of her possible motivations. Petrovna argues that the jealousy prompted by the situation led to the woman experiencing such extreme stress that she was not conscious or in control of her actions at the moment of castration: 'Complexes can exist independently, breaking away from the conscious control of the individual to act automatically.'[106] Brukhanskii similarly provided a rationale for the castration, rooting the crime in the woman's previous experiences and he also argued that at the moment the crime was committed Nastia E. was so overcome with emotion that she was incapable of acting rationally. Petrovna and Brukhanskii shared with Luria the conviction that violent crimes were associated with a 'primitive' psychic state, but Luria did less to explicitly root this regressive moment in the contemporary world nor did he analyse the class backgrounds or life experiences of the criminals he discussed. Luria was more concerned with assessing the emotional state of the criminal in the aftermath of the crime than at the moment in which it was committed.

Eric Naiman notes that Brukhanskii interpreted the case of Nastia E. and other cases of sexual violence he encountered in his clinical work 'as socially significant and linked to recent changes in Soviet social life', specifically to the kinds of shifts ushered in by the NEP.[107] Indeed, Naiman emphasises a broad trend for individual cases of sexual violence to be used to diagnose social pathologies in the period: 'The theme of the court as both mirror and diagnostic tool was a common motif in the NEP era.'[108]

[106] Petrovna, p. 100.

[107] Eric Naiman, *Sex in Public: The Incarnation of Early Soviet Ideology* (Princeton, NJ: Princeton University Press, 1997), p. 125. In a discussion of Brukhanskii's theorisations of crowd psychology Samuel Goff similarly claims that 'Violent or disorderly acts may be the outward manifestations of a kind of unregulated individualism [for Brukhanskii], but they can only be properly analysed relative to social living'. Samuel Goff, 'Physical Culture and the Embodied Soviet Subject, 1921–1939: Surveillance, Aesthetics, Spectatorship', PhD, University of Cambridge, 2017, p. 195.

[108] Naiman, p. 124.

Beer, Healey, Kowalsky and Kenneth Pinnow all cite examples of psychiatric cases discussing criminal acts that interpreted symptoms in relation to shifting Soviet social conditions, specifically as failures to adapt to the ideological mores of post-revolutionary society, but such interpretations remained only implicit in *The Nature of Human Conflicts.*[109]

In his sometimes floridly written introduction to the collection in which Petrovna's analysis is included, M.N. Gernet, who had established his credentials as a left-wing criminologist with his 1906 dissertation *The Social Causes of Crime*,[110] situated the various crimes discussed in the book—from serious incidents including mass murder to more mundane instances of petty theft and illicit alcohol production—to the social conditions of NEP-era Moscow. Though he delved into the pre-revolutionary history of crime in Moscow, compared the Soviet capital with other large European cities including Rome, Vienna and Paris, and attributed some of the preconditions of crime to modern city life with its 'bustling streets, crowded squares, high-rise buildings, busy theatres, cinemas, races, restaurants',[111] he also emphasised the particularities of post-revolutionary Moscow. In his typically bombastic rhetorical style, he wrote that in 1917, the year of the February and October Revolutions, 'Moscow was filled with desolate, hungry refugees ... giving children lessons in murder, robbery and violence' and noted that a prison amnesty resulted in the release of thousands of convicts who sought work in the city but often ended up on the streets.[112] He reported that between 1917 and 1923 though property crime decreased the murder rate in Moscow rose.[113]

Despite noting that people with 'psychopathic personalities' were active in the Moscow underworld, Gernet concluded by insisting on the importance

[109] See, for example, his discussion of a 1927 case by G.N. Udal'tsov, in which the clinical diagnosis of a formerly religious student is discussed in terms of his difficulties adjusting to the demands of the new regime (Beer, pp. 165–166). Healey similarly notes: 'Psychiatrists in their public statements [about rape cases] accepted the primacy of the economic and social determinants to the formation of human personality', p. 111. Kowalsky observes that criminologists 'relied on socioeconomic interpretations of crime', p. 16. And Pinnow writes that the 'the practice of social diagnostics expanded so dramatically in Russia during the 1920s', *Lost to the Collective*, p. 59.

[110] Kowalsky, pp. 38–39.

[111] M.N. Gernet, 'Predislovie' [Introduction], *Prestupnyi mir Moskvy* (Moscow: Pravo i zhizn', 1924), p. i.

[112] Gernet, p. xxvi.

[113] Gernet, p. xxvi.

of acknowledging the 'socio-economic conditions of crime',[114] taking into account factors that impact proletarian life including 'poverty, poor diet, bad education, alcoholism, syphilis, tuberculosis',[115] as well as the phenomenon of child homelessness (*besprizornost'*, which is discussed in more detail in the chapter on the figure of the child). He remarked that the articles in the collection were united by a 'red thread' in their attempts to show that even the most brutal crimes can be understood in the context of the brutal social world in which criminals have grown up and lived.[116] Contrasting the collection's approach with the Italian phrenologist Cesare Lombroso's assumptions about atavism and natural-born criminals, Gernet emphasised that crime should be understood in relation to historical and environmental conditions rather than as an a priori biological propensity.

A chapter on murderers in the book by S.A. Ukshe includes a discussion of the infamous case of Vasili Petrov-Komarov who was convicted of murdering 29 people between 1921 and 1923.[117] In a 1923 feuilleton Mikhail Bulgakov introduced the case in the following terms:

> Since the beginning of 1922 people had been disappearing in Moscow. For some reason, this happened most frequently to Muscovite horse traders and peasants from the villages near Moscow who came to buy horses. Apparently, the person did not buy a horse and then would disappear. At the same time, strange and unpleasant discoveries were turning up at night. In the empty lots of Zamoskvorechye, in ruined houses and in abandoned, unfinished bath houses on Shabolovka Street, stinking, grey sacks showed up. They contained naked male corpses. After several of these discoveries acute anxiety arose in the Moscow Criminal Investigation Department. The thing was, the sacks with the murdered men all bore the same fingerprints—it was the work of one man. The skulls were crushed with the same blunt instrument, the corpses were identically bound, always a neat and skilled job, with the

[114] Gernet, p. xli.

[115] Gernet, p. xl.

[116] Gernet, p. xli.

[117] The article gives the figure 29 though the press tended to report 33. See Mark Vincent, 'Making a Soviet Murderer: The Case of Moscow Serial Killer Vasili Petrov-Komarov', http://www.pushkinhouse.org/blog/2018/6/25/making-a-soviet-murderer-the-case-of-moscow-serial-killer-vasili-petrov-komarov. Vincent discusses another article dealing with the case by V. Larin in Mark Vincent, 'Cult of the "Urka": Criminal Subculture in the Gulag, 1924–1953', PhD, University of East Anglia, 2015, pp. 90–92.

> arms and legs drawn up to the stomach. The binding was done firmly, with care.[118]

Ukshe relates that the murders all followed the same pattern: the perpetrator would invite someone up to his apartment to discuss buying a horse, would ply them with vodka and hand them a document to peruse. The murders were committed by hitting the victims with a hammer covered with a tablecloth before strangling them with a noose. The corpses were then undressed, tied up and disposed of.[119] Although the short summary of the murders recalls those included in *The Nature of Human Conflicts* in its brevity and dispassionate descriptive style, as in the case of Nastia E., it is immediately supplemented by a discussion of the accused's upbringing and social situation. Interestingly, Ukshe's account emphasised a hereditary element to the criminal personality, drawing heavily on the work of Italian criminologist and student of Lombroso, Enrico Ferri:

> According to Ferri's definition, a premeditated murderer is a product of hereditary degeneration and suffers from an abnormality of physiological sensitivity. … This abnormality manifests as insensitivity to their own suffering and extreme dullness or even the lack of an instinct of self-preservation, as well as insensitivity to the suffering of their victim. Premeditated killers display cold cruelty during their crime and apathetic indifference after the fact, know no remorse, and describe the details of their murder with cynical composure.[120]

Ukshe's reasoning is not purely biological, however, as it linked the murderer's psychic state to the alcoholism of his parents, which thereby leads back to a socially grounded explanation. In his introduction Gernet described the murderer as a 'man-machine', echoing Bulgakov's account, which emphasised the killer's methodical approach, likening him to a dispassionate slaughterhouse worker: 'He killed neatly and with exceptional economy: always with the same instrument, with one blow to the skull, without noise or haste.'[121] Luria's experiments suggest that murderers can be identified through physical or verbal manifestations of guilt and

[118] Mikhail Bulgakov, 'The Komarov Case', *Notes on the Cuff and Other Stories*, trans. by Alison Rice (Ann Arbor, MI: Ardis, 1991), pp. 195–206, p. 195.

[119] S.A. Ukshe, 'Ubiitsy' [Murders], *Prestupnyi mir Moskvy*, pp. 41–81, p. 55.

[120] Ukshe, p. 47.

[121] Bulgakov, 'Komarov Case', p. 198.

assumed that murder necessarily stirs up emotion, whereas the profile of Petrov-Komarov insists on his calmness, lack of remorse and indifference, implying that a technique focused on the agitation of the suspect might not be universally applicable.

According to Ukshe few murders follow the pattern of Petrov-Komarov, as most are spontaneous rather than premeditated. Such cases, which more closely resemble those discussed by Luria, are instead characterised by a temporary emotional and moral disturbance resulting from 'feelings of anger, jealousy, revenge, fright, terror, pain, hunger, etc.'[122] Ukshe emphasised that these 'psycho-physical' reactions were tied to socio-economic conditions including 'unemployment, need, poor housing conditions and in particular such social disasters as crop failure, war, epidemics'.[123] As in the case of Nastia E. discussed by Petrovna and Brukhanskii, the cases of unpremeditated murder discussed by Ukshe include details about the perpetrator's social background and tend to contain a short description of the events immediately preceding the crime, which situate the violent acts in a particular context.

The litany of emotions Ukshe provided also stands in contrast to Luria's treatment of emotion which bundled different kinds of emotional responses together under the singular rubric of affect. In discussing a word association experiment with one of the men accused of murder, for instance, Luria described a range of responses that were observed—including stupor, tension, excitation and vigorous denial—yet he defines all these varied reactions as iterations of 'acute affect'.[124] Unusually in this case he also includes a description of the accused man's behaviour in the laboratory—'wiping his forehead, fidgeting about in his chair, and in the intervals between stimuli, he smacked his lips and muttered something to himself; from time to time there was a quick movement of the feet, and the speech responses were now very deep, almost whispers, now in a high-pitched tone of voice.'[125] But this plethora of actions, like the plethora of emotional states he indicated accompanied them, are subsumed under a monolithic concept. Ukshe was more interested in the social causes than the individual effects of crime and so distinguished jealousy from anger

[122] Ukshe, p. 69.

[123] Ukshe, p. 69. In his opening remarks the author similarly attributed any peak in the murder rate to the effects of the First World War, Civil War and famine, p. 45.

[124] Luria, *Nature*, p. 81.

[125] Luria, *Nature*, p. 90.

from lust from fear, whereas Luria in his laboratory reduced effects to the singular affect and disconnected both from particular experiences in a social world in the process.

Vygotsky expressed similar reservations about the word association method in a letter to Luria in 1926. He began by praising Luria's research (in relation to the purge experiments):

> The reactions accompanying the main process and side effects (pressure) reflect the outcome of the main process. Even what relates to it becomes more difficult, prolonged, destroyed. You now have proven that with different material. That is a big building block for the foundation of your previous studies, a confirmation of your methodology: It means that we can detect the fate of the main process (the Jungian as well as any other) through 'motor functions' and that we are entitled to interpret them as if they were symptoms. For me that is the most important aspect of your work; it tripled my belief in your previous work.[126]

He immediately followed these laudatory remarks by noting the possible limits of these experiments, however, specifically in relation to their capacity to record affect:

> The primary question for me, namely the question of methodology, is for me a question of truth, that is, of scientific discovery and imagination. However, theoretically I see great risks in the new experiments for your assumed conclusions so far: The line between affective and other damages fades away; the specific quality of the affect vanishes and your theory of emotions begins to crumble.[127]

By reducing affect to physical responses and by bunching all affects together, Luria risked treating emotion as quantitative rather than qualitative. Although pouring any liquid into a measuring jug makes it possible to find out its volume in millilitres, doing so would tell you nothing about its substance—milk, blood, wine, urine and lemonade would all be reduced to a number. Similarly, Luria's method treated anger, jealousy, fear, happiness and arousal as interchangeable. Guilt is connected to how much affect

[126] Vygotsky to Luria, Zakharino Sanatorium March 5, 1926, Georg Rückriem, ed., *Lev Lemenovic Vygotskij: Briefe/Letters, 1924–1934* (Berlin: Lehman, 2009), p. 216

[127] Rückriem, p. 216.

is displayed but the *content* of feeling remains unexamined.[128] Vygotsky's criticisms of Luria's theory of emotions recall Luria's own later critiques of Kornilov's definition of energy as overly mechanistic. Luria himself expressed reservations about the limitation of his methods to his English translator in America in 1931, indicating that he had since adopted different approaches: 'I am very glad that you like my book. I myself find in it a lot of flaws, and I can only regret that the schemes that served as the basis for it still do not adequately reflect the actual mechanisms of human neurodynamics.'[129] Indeed, even though most of the experiments included in *The Nature of Human Conflicts* were conducted after he first met Vygotsky in 1924, the shift in approach that followed could be attributed to the latter's increasing influence on Luria: 'I divide my whole biography into two periods: a brief, nonessential period preceded my meeting with Vygotsky, and a major and essential period, after I met him.'[130] In *The Nature of Human Conflicts* Luria also occasionally conceded that the quantification of affect his experiments relied on was reductive—'Our subject, however, is complicated and the living, human personality is not always adequately expressed in simple figures'[131]—but nowhere in the discussions of his experiments with criminals does he suggest this might invalidate his methodology which clings to the conviction that truth could be scientifically distinguished from lies.

Truth and Lies

Luria is sometimes credited with having invented the lie detector in connection with his experiments with criminals.[132] He even made this claim himself, implying that his techniques were a direct influence on subsequent

[128] Jacques Derrida's insights about Freud's economic conception of the relation between pleasure and unpleasure in 'Beyond the Pleasure Principle' are apposite here. Freud, he says, implicitly supposes that we know what pleasure is but that he fails to actually tell us anything about it—'Nothing is said of the qualitative experience of pleasure itself. What is it? What does it consist of? … The definition of the pleasure principle is mute about pleasure, about its essence and quality. Guided by the economic point of view, this definition concerns only quantitative relations.' Jacques Derrida, *The Postcard: From Socrates to Freud and Beyond* (Chicago, IL: University of Chicago Press, 1987), p. 276.

[129] Luria to W.H. Gantt, August 17, 1931, Alan Mason Chesney Archives, Johns Hopkins University, WH Gantt Papers, B45 F28.

[130] Luria, *Making of Mind*, p. 73.

[131] Luria, *Nature of Human Conflicts*, p. 111.

[132] T.V. Akhutina, 'A.R. Luriia: zhiznennyi put'' [A.R. Luria: The Way of Life], *Kul'turno-istoricheskaia psikhologiia*, 2 (2012), 2–10, p. 3. Kozulin speculates that Soviet lie detectors

devices in the USA but without substantiating the claim: 'the Americans … frivolously seized upon the idea, did not understand what it was all about and therefore misconstrued it'.[133] American lie detectors were the contemporaries of Luria's experiments in Moscow, so the simultaneous emergence of similar devices and methods in these two ideologically opposed nations within a juridical setting seems purely coincidental, at most indicative of the global reach of particular psychological technologies and approaches. Although a letter from the translator of *The Nature of Human Conflicts* to Luria mentions that people had written to him wanting to replicate the tests and asks technical questions about measuring muscle tension, the introduction of machines for measuring blood pressure to solve minor cases of petty theft in University of California sororities in the early 1920s preceded the American publication of *The Nature of Human Conflicts* by a decade.[134] Luria may not have been solely responsible for inventing the lie detector, but his use of machines to uncover truth does imply a particular understanding of truth and lies, demonstrating a faith in the physical detectability of the latter.

Though *The Nature of Human Conflicts* was not published in Russian in Luria's lifetime, Luria did publish some shorter pieces based on his experiments with criminals in the Soviet Union in the 1920s. Although it contained no details from particular cases or any case histories, Luria's article 'Experimental Psychology in Forensic Investigations' (1927) discussed the issues associated with attempts to establish guilt using psychological tests, declaring confidently that 'psychology, which studies the laws and forms of human behavior, will be able to play a significant role in judicial investigations in the future'.[135] The article's focus was on how to establish the truth of witness testimonies. Luria acknowledged that memory is complicated, slippery and sometimes unreliable. If a group of people

were based on Luria's experiments rather than making claims about American variants, *Psychology in Utopia*, p. 87.

[133] Levitin, p. 71. Here Luria also claimed his method was not adopted in the USSR but that there had been renewed interest in it in the late 1960s and early 1970s.

[134] Gantt to Luria, April 22, 1933, Alan Mason Chesney Archives, Johns Hopkins University, WH Gantt Papers, B45 F28. On the lie detector in America, see: Ken Adler, *The Lie Detectors: The History of an American Obsession* (New York, NY: Free Press, 2007).

[135] A.R. Luriia, 'Experimental psychology in forensic investigations' [Eksperimentantal'naia psikhologiia v sudebno sledstvennom dele], *Sovetskoe Pravo* [*Soviet Law*], 2, 26 (1927), 84–100, p. 85. The other article published in Russian discussing these experiments is: 'Psikhologiia v opredelenii sledov prestuplenii' [Psychology in the Investigatory Determination of Crime], *Nauchnoe Slovo*, 3 (1928), 77–104.

were asked to describe Theatre Square in Moscow or Raphael's Sistine Madonna, for instance, they would all emphasise different aspects and thus give slightly different accounts. Furthermore, he observed that people's responses to questions are effected by the conditions under which they are interrogated and by the educational or 'cultural' level of the respondent. Luria discussed the role of the interrogator, noting that the way their questions were formulated would influence the outcome of an interrogation. For example, asking a general question about a picture would result in a very different response from asking a leading question, such as: 'Was there a red flower in the picture?'[136] The latter contains an 'indirect suggestion',[137] which makes it more likely for the respondent to answer affirmatively, whereas the former is more objective. He ends by talking about the possibility of analysing the psychic life of investigators and judges themselves, though claims this is yet to be explored. Overall, in spite of his optimistic opening declarations, Luria is fairly modest about the applicability of the methods discussed in the article, underlining their limitations and acknowledging the complexity of disentangling truth from lies.

Luria's article cites a piece by Freud which engages with Jung's methods. Based on a lecture delivered shortly after Freud first began to correspond with Jung in 1906, 'Psychoanalysis and the Establishment of the Facts in Legal Proceedings' discussed the word association tests developed in Zurich. Following his rift with Jung, Freud was dismissive of these experiments,[138] and this earlier speech already expressed reservations about the applicability of the method. For Freud the untrustworthiness of witness statements made devising methods for ascertaining guilt that do not rely on verbal confession self-evidently appealing. He described the word association technique and praised Jung for establishing that the reaction to a stimulus word will never be accidental but will always reveal something about the test-subject's mental state. Freud linked this to his own insights in *The Psychopathology of Everyday Life* (1901) regarding verbal slips, memory lapses and lost objects, which similarly 'contributed towards restricting the arbitrary factor in psychology' by revealing the hidden

[136] Luria, 'Experimental Psychology…', p. 89.

[137] Luria, 'Experimental Psychology…', p. 89.

[138] Sigmund Freud, 'On the History of the Psychoanalytic Movement (1914)' in *Standard Edition of the Complete Works of Sigmund Freud*, vol. 14, trans. and ed. by James Strachey (London: Hogarth Press, 1957), pp. 7–66.

meaning lurking behind seemingly meaningless actions and statements.[139] He drew an analogy between the 'self-betrayal' associated with the revelation of 'complexes' through disturbed reactions to word association tests and the psychoanalytic method in cases of hysteria, and thus drew a parallel between the criminal and the hysteric. Though both criminal and hysteric have a secret they wish to keep hidden, Freud claimed that the difference between the two figures is that the criminal knows they are hiding something, whereas the hysteric does not, due to repression. Despite this key difference, however, Freud insists that the psychoanalyst proceeds just like a magistrate because their role is to 'uncover the hidden psychical material'.[140] Free association shares with word association the insistence that apparently arbitrary utterances, hesitations or mistakes betray something of psychological consequence.

Freud repeated that the main distinction between psychoanalysis and a criminal interrogation is primarily a distinction between the two people being examined or at least between their distinct motivations for undergoing the process of examination: 'In the neurotic the secret is hidden from his own consciousness; in the criminal it is hidden only from you.'[141] The neurotic is seeking to find a cure whereas the criminal is seeking to avoid arrest. The analysand is therefore more likely to be cooperative in the process than the criminal who is not seeking to 'combat his resistance' and whose resistance is conscious rather than unconscious.[142] Freud insisted that the distinction between unconscious and conscious resistance needs to be more carefully understood because the former is associated with unpleasurable experiences whereas the conscious withholding of information by a criminal may be connected with pleasure. A further complication arises because the neurotic and the criminal are not necessarily distinct; a neurotic accused of a crime may react guiltily 'because a lurking sense of guilt that already exists in him seizes upon the accusation made'.[143] Freud was concerned that distinguishing the person guilty of a specific crime

[139] Sigmund Freud, 'Psychoanalysis and the Establishment of the Facts in Legal Proceedings (1906)' in *Standard Edition of the Complete Works of Sigmund Freud*, vol. 9, trans. and ed. by James Strachey (London: Hogarth Press, 1959), p. 105.

[140] Freud, 'Psychoanalysis and Legal Proceedings', p. 108.

[141] Freud, 'Psychoanalysis and Legal Proceedings', p. 111.

[142] Freud, 'Psychoanalysis and Legal Proceedings', p. 112.

[143] Freud, 'Psychoanalysis and Legal Proceedings', p. 113. For a useful discussion of the psychoanalytic concept of denial and its implications for historians, see Catherine Hall and Daniel Pick, 'Thinking About Denial', *History Workshop Journal*, 84 (2017), 1–23.

from the person infused with a generalised feeling of guilt may not be as straightforward as the word association tests supposed.

Freud drew an analogy between psychoanalyst and magistrate, neurotic and criminal, while insisting on certain key distinctions between them, whereas Luria explored a situation in which these roles converged. He suggested that the criminal-neurotic may not only be consciously withholding information from the doctor-interrogator but that an unconscious element might also be observed. Luria claimed that the criminal may remember something significant about their crime but immediately force the memory back into their unconscious. In such cases they may even vigorously deny that an event even took place but the emotional way in which they express this negation would nonetheless alert the doctor to the lie. Luria claimed that the telling details that emerged in accounts of deeds intended to mislead interrogator's function like dreams or other forms of parapraxis in their capacity to reveal hidden meanings. He speculated that adding the analysis of such unconscious phenomena, 'applying the experience of psychoanalysis to judicial investigative practice', could eventually provide a new tool for establishing the distinction between truth and lies.[144] The unconscious with its capacity to simultaneously contain numerous contradictory 'truths' side by side is divested by Luria of its conflictual and ambiguous character.

Jung's discussions of word association tests indicate that he sometimes combined these methods with other forms of psychoanalytic inquiry that sought to uncover motives and delve into the life histories and upbringings of his patients. *Studies in Word Association*, for instance, relayed the case history of a woman who came to Jung for treatment for problems with sleeping. Tormented by anxious thoughts, she worried that she could 'infect' others with them. Jung noted that 'she complained of inner unrest and excitement'.[145] He described beginning the treatment with a series of word associations. These led him to identify what he assumed to be erotic disturbances, which he then used as the basis to begin a more conventional psychoanalytic session, relying on the Freudian method of free association. In psychoanalysis the patient related the story of a gardener who tried to kiss her. Even though she pushed the gardener away, she later fantasised about him coming into her room and sleeping with her. In another session she spoke about a servant who told her of her sexual

[144] Luria, 'Experimental Psychology…', p. 93.
[145] Jung, *Studies in Word Association*, p. 300.

experiences, which kept the patient awake at night as she reproached herself for feeling aroused. In this case the word association tests were used to reveal the 'complex of a psychogenic neurosis',[146] which psychoanalysis could then unravel and interpret; word association functioned as a means rather than an end. There is no indication that Luria employed this more detailed mode of analysis, however, though doing so may have enabled him to probe his interlocutors' motivations rather than merely establishing the fact of their guilt. By introducing free association into the process 'truth', in the example described by Jung, became more textured, multi-layered and contradictory rather than being reducible to a binary yes or no, which is how it remained in Luria's experiments with criminals.

Indeed, his experiments analysing affect in the real 'traumatic situations' of the examination and the aftermath of arrest seem to have led Luria in the opposite direction methodologically. Following his discussion of cases with criminals in *The Nature of Human Conflicts*, Luria bemoaned that the psychologist, unlike the chemist, must deal with factors that cannot be controlled or easily recorded, which led him not to delve into those complicating factors but to attempt to exclude them by turning to a discussion of experiments seeking to analyse affect by simulating responses under laboratory conditions: 'The ideal for the psychological experimenter has become the possibility to reconstruct artificially the phenomenon under examination, because only this enables one to keep it entirely under control.'[147] This paradoxically implies that the 'truth' of affect can best be revealed by artificial means. Under hypnosis a woman was told that she had been offered money to perform an illegal abortion which, despite her resistance to the suggestion, the doctor insisted she perform: 'her face changes, she

[146] Jung, p. 321.

[147] Luria, *Nature*, p. 129. A detailed discussion of the experiments in hypnosis and suggestion included in the book is beyond the scope of my discussion here. Luria's primary collaborator for these was Boris Efimovich Varshava who was also a member of the Russian Psychoanalytic Society and who died in 1927. K.I. Platonov, a major figure in Soviet psychotherapy, mentions collaborating with Luria in the early 1930s and Luria recommended an early article by Platonov for publication by an American journal (Hames, p. 169). Hypnosis and suggestion remained therapeutic methods in the Soviet psy-ences and hypnosis was approved of by Pavlov. See, for example, K.I. Platonov, *The Word as a Physiological and Therapeutic Factor: The Theory and Practice of Psychotherapy According to I. P. Pavlov* (Moscow: Foreign Languages Publishing House, 1959), Aleksandra Brokman 'Sterility and Suggestion: Minor Psychotherapy in the Soviet Union, 1956–1985', *History of the Human Sciences*, 31, 4 (2018), 83–106 and Wolf Lauterbach, *Soviet Psychotherapy* (Oxford: Pergamon, 1984).

trembles, makes restless movements on the couch and she is ready to cry.'[148] Though this situation was fictional, the emotion evoked was real, which seems to further complicate the suppositions about truth underlying Luria's experiments with criminals rather than bearing them out.

When Luria and his team set up their laboratory in Moscow they were assigned Lev Sheinin as an assistant 'at that time a young investigator of special importance'.[149] Sheinin, who was a journalist and successful playwright, also worked as an People's Commissariat for Internal Affairs (NKVD) official and went on to play a major role as an investigator in the Metro-Vickers case in 1933 and then the Moscow Show Trials of 1936–1938. It has been speculated that Luria's connection with Sheinin, who was part of state prosecutor Andrey Yanuaryevich Vyshinsky's 'friendly team' responsible for directing the trials, may have played a part in shielding Luria during the purges of the late 1930s.[150] According to Julie A. Cassiday, 'Sheinin played an exceptional role in creating both the show trials of the Great Terror and the fictional reality needed to back up their charges'.[151] Cassiday also notes that Sheinin's fictional works as a literary writer were informed by his 'experience as an investigator in the procurator's office of the RSFSR'.[152] Luria's connection with Sheinin does not necessarily imply any link between the experiments they conducted together and the approach to confession taken in the show trials, however; the mid-1920s should not be read through the prism of 1937.[153] Indeed, Luria's emphasis on the transparent relationship between truth and the

[148] Luria, *Nature*, p. 132. Interestingly in these artificially constructed situations Luria claims he did use the method of 'free association' referring to his method as 'experimental psychoanalysis', p. 151.

[149] Levitin, p. 70.

[150] Anton Yasnitsky and René van der Veer eds., *Revisionist Revolution in Vygotsky Studies* (London: Routledge, 2015), p. 46. Luria's father also served as a Kremlin doctor and his role in treating Vyshinsky's gastronomic complaints may also have been a factor in protecting his son.

[151] Julie A. Cassiday, *The Enemy on Trial: Early Soviet Courts on Stage and Screen* (DeKalb, IL: Northern Illinois Press, 2000), p. 194. On the status of confession in agitation trials in the 1920s, see Elizabeth A. Wood, *Performing Justice: Agitation Trials in Early Soviet Russia* (Ithaca, NY: Cornell University Press, 2005).

[152] Cassiday, p. 193.

[153] On shifts in Soviet law and policing in this period, see Paul Hagenloh, *Stalin's Police: Public Order and Mass Repression in the USSR, 1926–1941* (Baltimore, MD: Johns Hopkins University Press, 2009) and Louise I. Shelley, *Policing Soviet Society: The Evolution of State Control* (London: Routledge, 1996).

body, the incapacity for people to hide the truth from the interrogator intent on uncovering it, seems almost the inverse of the kinds of forced confessions based on fabricated accusations often extracted under torture that would come to characterise the show trials. If Luria's methods focused on extracting truth from lies and assumed confession would result in a cathartic release of tension and conflict produced by lying,[154] Sheinin was later involved in forcing people to lie and to confess to crimes they had not committed in order to provide 'proof' of guilt.[155] Though the example of the show trials indicates how pliable 'truth' can become if the temperature is turned up high enough.[156]

A more telling comparison can be found between Luria's accounts of his experiments (both in *The Nature of Human Conflicts* and in his subsequent autobiographical narratives) and Lev Sheinin's recollections of the same period. In autobiographical reflections published as an introduction to a collection of vignettes and short stories, Sheinin reflects on his early career in Moscow beginning in the year 1923, when Luria also first arrived in the city. He claims that his literary career 'developed at the interrogation table'.[157] Though he does not mention his involvement with Luria's experiments explicitly, he does stress the psychological dimension to the role of the investigator:

> whatever the case may be—whether it is about robbery, murder committed out of jealousy, or embezzlement and bribery—behind it always and, above all, there are people, each with their own character, their own destiny, their own feelings. Without understanding the psychology of these people, the investigator will not understand the crime they committed. Without understanding the inner world of each accused person, the complex, sometimes

[154] Luria, *Nature*, p. 114. Luria uses the term 'catharsis' in relation to confession in A.R. Luria, 'The New Method of Expressive Motor Reactions in Studying Affective Traces', *Proceedings and Papers of the Ninth International Congress of Psychology (held at Yale University, New Haven, Connecticut, September 1st–7th 1929)* (Princeton, NJ: Psychological Review Company, 1930), pp. 294–296, p. 295.

[155] Sarah Davies, *Popular Opinion in Stalin's Russia: Terror, Propaganda and Dissent, 1934–1941* (Cambridge: Cambridge University Press, 1997), pp. 113–123 and Sheila Fitzpatrick, *On Stalin's Team: The Years of Living Dangerously in Soviet Politics* (Princeton, NJ: Princeton University Press, 2015) (particularly Chap. 5).

[156] Luria may have explored suggestibility in other experiments in the period, but he drew a clear line between conscious and hypnotic states.

[157] Lev Sheinin, 'Rasskaz o cebe' [Story of Myself], *Zapiski sledovatelia* [*Notes of an investigator*] (Moscow: Sovetskii pisatel', 1968), pp. 5–20, p. 5.

> surprising combination of circumstances—accidents, vices, bad habits and connections, weaknesses and passions—the investigator will never understand the case.[158]

Here Sheinin invoked a litany of emotions and possible motivations that recalls Ukshe's discussion of unpremeditated murders. Sheinin also evocatively captured the atmosphere of NEP-era Moscow in a style closer to Gernet's introduction to *The Criminal World of Moscow* than to Luria's own publications:

> Moscow of my youth I will never forget you!—I close my eyes and see your snow-covered streets: the narrow Tverskaya with the chapel of the Iverskaya Mother of God in Okhotny Ryad, occasional moaning trams, sleepy cabbies at intersections, horses slowly chewing oats from hanging sacks, Mosselprom saleswomen … selling chocolate and 'Ira' cigarettes (which were said to be "all that remained of the old world"). … I see your stations, densely populated student dormitories, long cheerful night queues at the box office of the Moscow Arts Theatre.[159]

Employing tropes familiar from accounts of NEP-era Moscow, as a space of contradiction, where the new rubbed shoulders uncomfortably with the vestiges of the old,[160] Sheinin vividly recalled Komsomol clubs, bustling markets, drunken gatherings of imagist poets, anarchist groups, students of Esperanto hoping to hasten world revolution by mastering a global language as well as more nefarious characters: NEP men, cheats,

[158] Sheinin, pp. 8–9.

[159] Sheinin, p. 5.

[160] According to William G. Rosenberg, temporal contradictions were at the heart of NEP-era political policy: 'NEP was a period in which those in power were forced somehow to come to terms with complex social and cultural residues of pre-revolutionary Russia, implicitly at odds with ongoing goals of building socialist or communist order.' 'Introduction: NEP as a "Transitional" Society' in *Russia in the era of NEP: Explorations in Soviet Society and Culture*, ed. by Sheila Fitzpatrick, Alexander Rabinowitch and Richard Stites (Bloomington: Indiana University Press, 1991), pp. 1–11, p. 3. These kinds of juxtapositions can be found in first-hand accounts of foreign visitors to Moscow in the period. See, for example, Walter Benjamin, *Moscow Diary*, trans. by Gary Smith (Cambridge, MA: Harvard University Press, 1986) and Dorothy Thompson, *The New Russia* (New York, NY: H. Holt and Co., 1928). In his 1927 account of Soviet Russia René Fülöp-Miller, described 'a fantastic between-world … a medley of mixed forms chaotically jumbled together … the Asiatic past with the latest achievements of technology'. René Fülöp-Miller, *The Mind and Face of Bolshevism: An Examination of Cultural Life in Soviet Russia*, p. 215.

swindlers, crooks, speculators, sex workers, sex maniacs, 'gangsters with aristocratic manners and former aristocrats who had become gangsters'.[161] He described the new kinds of companies that sprung up in the mixed economy and the black-market trade in luxury goods. He was appointed as an investigator due to his membership of the Komsomol. Like Luria at the Institute of Psychology, youthful enthusiasm took the place of professional experience. These reminiscences of Moscow might be laced with literary flourishes and nostalgic invention but they conjure the textures of a world beyond the laboratory and interrogation room that situate criminality in a particular physical *and* discursive context. Discursive tropes about NEP-era Moscow may not have precisely mapped on to the empirical realities of life in the city, but that does not mean they did not participate in shaping how people understood and hence inhabited the world around them.

In contrast, by treating history as the centuries long development of civilisation from 'primitivism', in *The Nature of Human Conflicts* Luria had less interest in granular descriptions of the fabric of life that characterised the specific historical environment that the people he was analysing inhabited. The textures of the world in which he and the people he was experimenting on lived are thus given only a cursory gloss in Luria's accounts of his experiments and the social details that do poke through his prose, like plants through the cracks in the pavement, are incidental to his conclusions. For all his insistence on distinguishing truth from lies, 'truth' was not understood as a form of veracity with depth or qualities. Though not asking people to 'confess' to false accounts of their actions or motivations as in the later show trials, he was not therefore attempting to understand people's actions in relation to their specific experiences, to situate them in relation to broader structural conditions or to prevent similar acts from being repeated in the future.

Conclusion

The Nature of Human Conflicts is dedicated to the American philosopher Horace Kallen, who visited the Soviet Union in 1928, and recalled meeting Luria in his 'dingy room' in which the psychologist and his first wife (who had an affair and left him soon after) 'slept and cooked and ate and worked', preparing their meals on a primus stove whilst earning subsistence-

[161] Sheinin, p. 6.

level wages.[162] Kallen remarked upon the irony that those people upon 'whom avowedly the new Russia depends for a happy and prosperous future' were forced to live in such insalubrious conditions, reflecting that this gap between future promise and lived reality explained 'why whatever Moscow gayety one encountered seemed so forced'.[163] In Luria's theoretical works from this period the liberated future glows faintly but surely on the horizon, whereas *The Nature of Human Conflicts* remains with neurotic, 'disorganised' and disturbed individuals in the gloomy present but without analysing the shared social conditions that may have contributed to producing their violent actions or affective reactions in any systematic way.

For the most part *The Nature of Human Conflicts* lacks the overarching meta-historical framework that informed Luria's more explicit engagements with Freud. It therefore sidesteps the uncomfortable question of how to reconcile examples of regression and aggression with his framework for understanding the progressive movement of history. Instead, a similarly progressive model of development is transferred from the historical on to the individual, which allowed Luria to acknowledge regressions as isolated deviations from a normative developmental trajectory. Yet Luria's analysis of individual development had implications for his conceptualisation of historical development elsewhere, for the individuals he described, by his own oft-repeated definition of human psychology, were not external to history, as he stated unequivocally in 'Psychoanalysis as a System of Monisitc Psychology': 'Marxism sees the individual as an inseparable element of and an active force in history'.[164] The utopian assumptions that animated his and Vygotsky's discussions of Freud's *Beyond the Pleasure Principle*, which insisted that the instinctual conservatism Freud identified as an integral aspect of human nature could be overcome through radical social change, was undercut by the individual cases

[162] Horace Kallen, *Frontiers of Hope* (New York, NY: Arno Press, 1929), p. 289. Kallen does not mention Luria by name, referring only to a 'brilliant young scholar', but Luria makes reference to the passage in a letter to Kallen responding to Kallen's book. Luria praises Kallen's depiction of Moscow but adds wryly that he and his wife did, in fact, have a separate room in which to cook. See, Luria to Kallen, September 29, 1929, Horace Kallen Papers, Cincinnati, OH, American Jewish Archives (AJA), Box 19, Folder 18, Luria, Alexander R. 1929–1972.

[163] Kallen, *Frontiers*, p. 290.

[164] Luria, 'Psychoanalysis...', p. 32.

described in *The Nature of Human Conflicts*.[165] In their introduction to *Beyond the Pleasure Principle* Luria and Vygotsky concluded by remarking that '[b]ourgeois science is giving birth to materialism'.[166] In practice, however, *The Nature of Human Conflicts* suggests that Luria failed to overcome the gap between 'substance and spirit' that he identified with the bourgeois scientific approach.[167]

In 'Principle Questions of Contemporary Psychology' Luria argued that in order to achieve a holistic understanding of human behaviour, forms of experimentation needed to be devised that refused to artificially separate out different aspects of the personality and instead created a more 'natural' situation in which people could be observed interacting with their immediate environment. He gave as an example an experiment devised by German Gestalt psychologists based in Berlin in which a person is placed in a room empty but for a table and a small selection of objects and is then observed interacting with the space. Some subjects, he noted, concentrated on themselves while others interacted with the objects in the room. Although he notes this method could be improved upon by developing a more dialectical approach, he was nonetheless broadly approving, seeing the example as a nascent method for understanding the many facets of human personalities.[168] Yet it is striking that as a counterpoint to 'artificial' modes of experimentation, Luria proposed a technique in which a person was placed in an empty room in laboratory conditions and left to interact with objects that had no bearing on that person's everyday concerns. In the early 1930s, however, Luria led two expeditions to Soviet Central Asia which saw him depart from his laboratories to conduct experiments with people in their own social environments. Unlike the examples included in *The Nature of Human Conflicts*, in conceiving these experiments Luria was explicit in his attempt to ground people's cognitive processes not simply in history in the broadest sense, but in the specific social transformations then taking place in the region.

[165] Luria and Vygotsky, 'Introduction…', p. 16.
[166] Luria and Vygotsky, 'Introduction…', p. 17.
[167] Luria, 'Psychoanalysis…', p. 27.
[168] Luriia, 'Printsipial'nye voprosy…', pp. 137–139.

CHAPTER 3

The 'Primitive'

In 1931 and 1932 Alexander Luria led two psychological expeditions to Soviet Central Asia. He hoped to trace the changes in thought he assumed would accompany the social alterations wrought by the First Five Year Plan (1928–1932), the 'great break' ushered in by the Soviet Union's Cultural Revolution or 'second revolution'.[1] Luria hoped that this moment of rapid social upheaval would provide him with the material to support his theoretical insistence on the historical contingency of human thought. He described his experiments as

> a statement of the fundamental shifts that had occurred in human consciousness during a vigorous realignment of social history—the rapid realignment of a class society and a cultural upheaval creating hitherto unimagined perspectives for social development.[2]

The experiments set out to show that thought structures were tied to material conditions, framing psychology as the 'science of social history'.[3]

[1] On the term 'Cultural Revolution' (incidentally from a text published in the same year Luria published his findings from this expedition), see Sheila Fitzpatrick, 'Cultural Revolution in Russia 1928–1932', *Journal of Contemporary History*, 9, 1 (1974), 33–52. The latter term appears in Isaac Deutscher, *Stalin: A Political Biography* (London: Penguin, 1968), p. 296.

[2] A.R. Luria, *Cognitive Development: Its Cultural and Social Foundations*, trans. by Martin Lopez-Morillas and Lynn Solotaroff (Cambridge, MA: Harvard University Press, 1976), p. 19.

[3] Luria, *Cognitive Development*, p. 12.

H. Proctor, *Psychologies in Revolution*, Mental Health in Historical Perspective, https://doi.org/10.1007/978-3-030-35028-4_3

Moreover, Luria intended to demonstrate that increased literacy levels, new forms of collective labour and exposure to modern technology not only changed but improved people's thought processes, introducing the capacity for abstract thought and thus prompting an ascent to a more advanced cognitive stage:

> studying not only the peculiarities of psychological processes but, what is more important, the very dynamics of the transition from the more elementary psychological laws to the more complex processes … connected with socialistic growth.[4]

Vygotsky, who was involved in the conception of the project and conducted a series of preliminary experiments in 1929 but was too ill to travel with the group, heralded the study as an unprecedented investigation of 'historical phylogeny', in which a cultural transformation that would ordinarily occur over many generations could be witnessed on the ontogenetic level, within the lifetime of an individual.[5]

Luria's expeditions, supported by a range of state-run institutions, were part of a flow of experts from a range of disciplines who travelled from Russia's Western metropolitan centres to distant parts of the Soviet Union as emissaries of the revolution.[6] The Central Asian expeditions were perhaps Luria's most explicit political endeavour: an attempt to demonstrate

[4] A.R. Luria, 'The Second Psychological Expedition to Central Asia', *Journal of Genetic Psychology*, 44 (1934), 255–259, p. 255.

[5] Vygotsky to Luria, August 1, 1931, cited in Elena Luria, *Moi Otets: A.R. Luria* [*My Father: A.R. Luria*] (Moscow: Gnosis, 1994), p. 61. The historically unprecedented character of the experiments (and, implicitly, the First Five Year Plan) is echoed in letters Luria sent to Horace Kallen (Luria to Kallen, October 28, 1931, Cincinnati, OH, American Jewish Archives (AJA), Horace Kallen Papers, MS-1, Box 19, Folder 18) and W.H. Gantt (Luria to W.H. Gantt, August 17, 1931, Alan Mason Chesney Archives (AMCA), Johns Hopkins University, W.H. Gantt Papers, B45 F28.)

[6] On state-funded artistic and literary projects undertaken during the First Five Year Plan, see Maria Gough, 'Radical Tourism: Sergei Tret'iakov at the Communist Lighthouse', *October*, 118 (2006), 159–178 and Mieka Erley, '"The Dialectics of Nature in Kara-Kum": Andrei Platonov's *Dzhan* as the Environmental History of a Future Utopia', *Slavic Review*, 73, 4 (2014), 727–750. On 'communist missionaries' from other disciplines, see 'The Conquerors of Backwardness', Yuri Slezkine, *Arctic Mirrors: Russia and the Small Peoples of the North* (Ithaca, NY: Cornell University Press, 1994), pp. 187–300. Luria's expeditions were backed by the People's Commissariat for Education of the Uzbek Soviet Socialist Republic and the Government of Uzbekistan. See Luria, 'Second Psychological Expedition…', p. 255.

the cognitive benefits of collectivisation. Yet, despite his declared allegiance with the Stalinist project, Luria's work in Central Asia was denounced before he published his findings. David Joravsky characterises the attacks on Luria as part of the 'vituperative spirit of the 1930s'.[7] Luria's research papers were taken for examination by The Culture and Propaganda Department of the Central Committee of the Communist Party of the Soviet Union (*Kult'prop*) as part of a larger inspection of the Institute of Psychology in Moscow. Following the investigation, he was accused of 'extremely grave charges' including racism, colonialism and an overreliance on 'bourgeois' Western theories.[8] Luria reflected that this was part of a project to discredit him as a scientist committed to the Soviet cause. He wrote to the People's Commissariat for Education to plead his case, explaining that the judgements against him had been based on research in its 'raw form' and interpreted without any understanding of the methodologies or intentions of the project.[9] This was to no avail: the experiments were declared harmful to the construction of a national culture in Uzbekistan. In the words of one ideologue:

> this pseudo-scientific, reactionary, anti-Marxist and anti-class theory, in practice leads to the anti-Soviet conclusion that policy-makers in the Soviet Union are people and classes that think primitively and are incapable of any kind of abstract thinking, which, of course, does not at all fit the facts.[10]

As Raymond Bauer discussed in a 1952 survey of Soviet psychology, the accelerationist pace of the First Five Year Plan was accompanied by a new intolerance for any acknowledgement of a gap between the imperfect present and the Communist future. Although Luria was enthusiastically attempting to trace positive changes in consciousness, his experiments still

[7] David Joravsky, *Russian Psychology: A Critical History* (Oxford: Blackwell, 1989), p. 365. See also, Alexei Kozulin, *Psychology in Utopia: Toward a Social History of Soviet Psychology* (Cambridge, MA: MIT Press, 1984), p. 110.

[8] E. Luria, *Moi Otets*, p. 67.

[9] E. Luria, *Moi Otets*, p. 67.

[10] P. Razmyslov, 'On Vygotsky and Luria's "Cultural-Historical Theory of Psychology"', *Journal of Russian and East European Psychology*, 38, 6 (2000), 45–58, p. 54. Originally published as '"O Kulturno-istoricheskoi teorii" Vygotskogo i Luriia', *Kniga i proletarskaia revolutsiia*, 4 (1934), 82–84.

described communism as a work in progress, whereas psychologists' findings were now expected to describe the already realised ideal Soviet person.[11]

The eventual suppression of this research should not, however, obscure the extent to which Luria and his team were active and enthusiastic participants in the Soviet project. Indeed, despite describing the 'backwardness' of Uzbek people who had not yet been exposed to the new forms of life being ushered in by the First Five Year Plan, Luria insisted that dramatic changes in thought processes could already be observed in Central Asia. He declared the situation 'a genuine revolution in cognitive activity',[12] describing a psychic upheaval as transformative as the political shifts to which it was tied. Luria wrote excitedly to the American anthropologist Melville Herskovits in 1932:

> We have in the USSR extremely interesting and happy conditions to study the process of very deep and very quick changes in the psychological processes of the periphic [*sic*] nations. The process of collectivisation of economic forms and industrialisation, the socialistic reconstruction of life does involve real changes in the mind.[13]

Paradoxically, however, Luria conceived of the imposition of a particular mode of life and thought as a form of liberation. For Luria, the transition to abstract thinking did not represent a process of assimilation but of emancipation. Only by developing the capacity to think abstractly would Uzbek people be capable of actively shaping historical development; but if psychology, as Luria claimed, is the 'science of social history', then these utopian notions should be understood in the social and historical context in which his experiments were conducted.

In 1967 Luria informed Jerome Bruner that he planned to publish the results of his experiments conducted in the 1930s in Central Asia. He said that he conceived of the work as part of a 'romantic series' that would also

[11] See Raymond Bauer, *The New Man in Soviet Psychology* (Cambridge, MA: Harvard University Press, 1952), pp. 107–114. The shift in psychology that accompanied the onset of the First Five Year Plan had counterparts across disciplinary fields: 'The greatest change in scientific establishment of the Soviet Union occurred in the years of the Cultural Revolution, 1928–1931.' Loren Graham, *Science in Russia and the Soviet Union* (Cambridge: Cambridge University Press, 1993), p. 93.

[12] Luria, *Cognitive Development*, p. vi.

[13] Luria to Herskovits, April 19, 1932, Chicago, Northwestern University Archives (NUA), Melville J. Herskovits Papers, General, 1906–1942, Box 13, Folder 4, 1931–1935.

include his two case histories, *The Mind of a Mnemonist* and *The Man with a Shattered World*.[14] Luria contrasted the warm 'romantic' approach taken in these late publications to the cold and abstract 'classical' approach typical of most scientific writing, but the resulting book *Ob istoricheskom razvitii poznavatel'nykh protsessov* [*On the Historical Development of Cognitive Processes*] (published in 1974 and translated into English as *Cognitive Development: Its Cultural and Social Foundations* in 1976) ultimately has little in common either in style or in clinical content with those case histories, which each focus on an individual with whom Luria worked for many years. Indeed, it does not even include the kind of compressed thumbnail case histories included in *The Nature of Human Conflicts*. Instead, the book summarises responses to various tests involving geometrical shapes, colours and syllogisms, conducted with many people about whom the reader is informed of little other than their social position and educational level.[15] This chapter first attempts to provide a more detailed account of the social, cultural and historical situation in which Luria's experiments were conducted than he provides in the book, factors he claimed were central to its core arguments about psychology. It then concludes with close readings of transcripts of experiments included in the book, contending that despite the brevity of the responses recorded these protocols nonetheless betray evidence of resistance to Luria's methods and ultimately pull against his conclusions, gesturing towards a more complicated definition of revolutionary subjectivity.

Psychology as Social History

There is a marked dissonance between the time in which Luria wrote up his findings from the expeditions to Central Asia in the 1970s and the time in which he conducted the experiments. Following the expeditions, short bulletins appeared in contemporary English-language journals,[16] but due

[14] Luria to Bruner January 9, 1967, HUA, Jerome Bruner Papers, HUG 4242.5 General Correspondence 1964–1965. See also, Luria to Brian and Joan Simon May 16, 1973, IOE, Brian and Joan Simon Papers, DC/SIM/2/2. This letter gives the book's working title as *The History of Cognitive Processes* and describes it as 'the book dealing with the results of my expedition to Middle Asia in 1930–1931, showing the Cultural Revolution, liquidation of illiteracy, short schooling etc.'

[15] A.R. Luriia, *Ob istoricheskom razvitii poznavatel'nykh protsessov* [*On the Historical Development of Cognitive Processes*] (Moscow: Nauka, 1974).

[16] See Luria, 'Psychological expedition to Central Asia', *Science*, 74 (1931), 383–384, Luria, 'Psychological

to the suppression of this research in the Soviet Union, a book-length account of the experiments did not emerge for more than 40 years. Luria concluded that text with the pronouncement: 'In the past forty years, a backward remote region has become an economically and socially developed part of our socialist state'.[17] The political moment of Brezhnev's stagnation was a stark contrast to the years of rapid and violent change ushered in by the First Five Year Plan, during which the experiments were conducted.

The accounts of Gestalt psychologist Kurt Koffka, who accompanied the second expedition to Central Asia in 1932, captured some of the enthusiasm he encountered in the Soviet Union. In Moscow, Koffka claims that the atmosphere among the Soviet psychologists was 'extremely lively and optimistic. One seems to feel the joint effort of everybody proud of what has been achieved, and is being done, now, and planned for the future.'[18] Koffka's notes on the 1932 expedition bluntly state that he was informed by an Uzbek representative in Moscow: 'Nothing achieved by force'.[19] It is, however, unlikely that Luria and his colleagues were oblivious to the brutal means by which collectivisation was achieved or to its outcomes, even if they did not necessarily witness its most extreme effects first-hand.[20]

Uwe Gielen and Samvel Jeshmaridian have attacked Luria's experiments in Central Asia for their complicity with Stalinism, insisting on the

Expedition to Central Asia', *Journal of Genetic Psychology*, 40, 1 (1932), 241–242, Luria, 'The Second Psychological Expedition to Central Asia', *Science*, 78 (1933), 191–192 and Luria, 'The Second Psychological Expedition to Central Asia', *Journal of Genetic Psychology*, 44 (1934), 255–259.

[17] Luria, *Cognitive Development*, p. 164.

[18] Molly Harrower, *Kurt Koffka: An Unwitting Self-Portrait* (Gainesville, FL: University Presses of Florida, 1983), p. 146. Harrower was Koffka's research assistant. This volume includes their correspondence during his time in Central Asia. Harrower continued to correspond with Luria about her own research after Koffka's death in 1941.

[19] Akron, Ohio, Archives of the History of American Psychology (AHAP), Kurt Koffka Papers, Box M379, 'Lectures Misc'.

[20] Tatiana Akhutina, a former student and collaborator of Luria's, made this observation when I presented an early version of this material at the International Society for Cultural and Activity Research's fourth International Summer University in Povedniki, Russian Federation, July 2013. It should also be remembered that Luria was himself living in Kharkiv in Ukraine in the early 1930s. Vygotsky's daughter recalled being afraid of relocating to Kharkiv as she had heard adults talking about the famine in the region. See GL Vygodskaia and TM Lifanova, 'Lev Semenovich Vygotsky (Part 3)', *Journal of Russian and East European Psychology*, 37, 4 (1999), 3–86.

importance of acknowledging that Luria and Vygotsky 'lived in an era when millions of people starved to death, were sent to labour camps, were murdered'.[21] In advancing this argument the authors cite statistics from a single source focused on Ukraine: Robert Conquest's *The Harvest of Sorrow* (1986), which was sponsored by the Harvard Ukrainian Research Institute and the Ukrainian National Association.[22] This book not only focuses on a region of the Soviet Union other than Uzbekistan[23] but was also published before the subsequent 'archive revolution'. Glen and Jeshmaridian's article relies on juxtaposing decontextualised numbers and harrowing descriptions of the famine induced by collectivisation in Ukraine and Kazakhstan with explicitly ideological statements by Luria and Vygotsky.

[21] Uwe Gielen and Samvel Jeshmaridian, 'Lev Vygotsky: The Man and the Era', *International Journal of Group Tensions*, 28, 3/4 (1999), 273–301, p. 276. Though less condemnatory, a similarly bemused account of Luria's relation to collectivisation which also relies primarily on Conquest is given in René van der Veer and Jaan Valsiner, *Understanding Vygotsky: A Quest for Synthesis* (Oxford: Blackwell, 1991), pp. 243–246.

[22] Robert Conquest, *The Harvest of Sorrow: Soviet Collectivization and the Terror-Famine* (Oxford: Oxford University Press, 1986). On Conquest generally and *Harvest of Sorrow* specifically, see Sheila Fitzpatrick, 'People and Martians', *London Review of Books*, 41, 2 (2019), 13–15. Other major historical accounts of collectivisation include: R.W. Davies and Stephen G. Wheatcroft, *The Industrialisation of Soviet Russia Volume 5: The Years of Hunger: Soviet Agriculture 1931–1933* (New York, NY: Palgrave Macmillan, 2004), Sheila Fitzpatrick, *Stalin's Peasants: Resistance and Survival in the Russian Village After Collectivization* (Oxford: Oxford University Press, 1994), Moshe Lewin, *Russian Peasants and Soviet Power: A Study of Collectivisation* (New York, NY and London: W.W. Norton, 1975) and Lynne Viola, *Peasant Rebels under Stalin: Collectivisation and the Culture of Peasant Resistance* (Oxford: Oxford University Press, 1996).

[23] For discussions of collectivisation in the Uzbek SSR specifically, see R. Aminova, *Changes in Uzbekistan's Agriculture, 1917–1929*, trans. by V.A. Epshtein and B.N. Lunkov (Moscow: Nauka Publishing House, 1974), Shoshana Keller, *To Moscow, Not Mecca: The Soviet Campaign Against Islam in Central Asia, 1917–1941* (Westport, CT: Praeger Publishers, 2001), pp. 175–211 and Michael Rywkin, *Moscow's Muslim Challenge: Soviet Central Asia* (Armonk, NY: M. E. Sharpe, 1982). A useful summary is provided by Marianne Kamp: 'In the first years of collective farms (*kolkhozes*), living standards plummeted for most rural people; those who were even moderately well-off lost their land and livestock to the collective farms, and many farms were not allowed to raise necessary food crops, but instead had to plant cotton. In the early 1930s, more than thirty-one thousand households were dispossessed and their members exiled, either through forced resettlement in Uzbekistan or to labor colonies elsewhere in the Soviet Union. This reorganization of the countryside greatly expanded the state's ability to control and intervene in the everyday life of most Uzbeks.' Marianne Kamp, *The New Woman in Uzbekistan: Islam, Modernity, and Unveiling under Communism* (Seattle, WA: University of Washington Press, 2006), p. 218.

The authors are correct to observe that the psychologists' relation to Soviet history had been overlooked by most Western secondary literature. However, by reducing the nuances and textures of that history to the worst excesses of Stalinism, they do little to illuminate the nature of that relation. Instead, their analysis imagines Stalinism as a phenomenon so extreme and monolithic that it defies human comprehension, which thus seals it off from proper scrutiny. Moshe Lewin identifies the necessity to confront contradiction as one of the fundamental problems faced by historians of the Stalin era:

> That a progressive ideology, initially intended to enhance human freedom and to create higher forms of community, came to serve a police state is one of the peculiarities of the period and an important phenomenon to study.[24]

This is not a comfortable endeavour but, to make sense of Luria's project and its relation to Stalinism, as Alexei Kozulin observes, it is important to take his commitment to the Soviet cause seriously.[25] Susan Reid has convincingly demonstrated in the field of the visual arts that Western narratives of Soviet intellectuals as a 'browbeaten bunch' meekly kowtowing to the Soviet authorities fail to account for the animated and committed attempts to engage with and implement state policy in different fields, including by those who were subsequently persecuted.[26] Neither is it helpful to suggest, as Gielen and Jeshmaridian risk doing, that articulating support for collectivisation and mass literacy campaigns is simply equivalent to the advocation of mass murder or apologism for famine. Furthermore, Gielen and Jeshmaridian present Luria's positive proclamations about the First Five Year Plan as evidence of his complicity with Stalinism without examining in any detail the methodological approaches he adopted in Central Asia.

Luria is alleged to have telegrammed Vygotsky from Samarkand saying simply: 'The Uzbeks have no illusions!' The story goes that this message, which Luria intended as a summary of his research findings, was intercepted by Soviet authorities and interpreted as a condemnation of collec-

[24] Moshe Lewin, *The Making of the Soviet System: Essays in the Social History of Interwar Russia* (London: Methuen, 1985), p. 210.

[25] 'One cannot do justice to Luria and his generation without recognising their sincerity.' Kozulin, p. 15.

[26] Susan E. Reid, 'Socialist Realism in the Stalinist Terror: The Industry of Socialism Art Exhibition', 1935–1941, *Russian Review*, 60, 2 (2001), 153–184, p. 161.

tivisation.[27] Regardless of the veracity of this anecdote, the statement does encapsulate how Luria characterised Uzbek thought. Ironically, the lack of illusions that Luria identified in Uzbek people was precisely what he hoped could be overcome through the transition to the socialism. For Luria, illusion, fantasy and imagination were all crucial components of 'advanced' thinking. In a letter to the Gestalt psychologist Wolfgang Köhler in the wake of the 1931 expedition he noted: 'illusions seem to be a function of a culturally highly developed psyche'.[28]

Luria saw the development of 'higher' forms of thought as one of the most fundamental psychological questions. In Central Asia, Luria intended to trace the transition from concrete to abstract thought which he assumed would accompany the social and economic changes taking place. This transformation was for him indicative of cultural progress, as he assumed that 'a person capable of abstract thought reflects the external world more profoundly and completely'.[29] Luria's interest in abstract thinking was shared by numerous psychologists of the period. Abstraction, similarly defined as the ability to classify and order information into predetermined categories, was also central to the work of the German neurologist Kurt Goldstein, for example. Anne Harrington argues that Goldstein's discussions of the capacity for abstract thinking in the 1920s and early 1930s, which he viewed as integral to human freedom, were imbued with a particular culturally specific political meaning: 'a reflection of his own liberal hopes for Germany'.[30] In Luria's work abstract thinking was also linked to freedom, and the two authors shared a connection to the German idealist and the romantic traditions discussed by Harrington, but Luria did not couch his discussions in the liberal rhetoric of choice favoured by Goldstein.

[27] There are various versions of this anecdote in circulation. Elena Luria quotes a lecture given by her father in 1974 that refers to this incident without, however, mentioning any political repercussions. In an oblique footnote, she adds: 'This telegram, which was deliberately misinterpreted by their enemies, got Luria and Vygotsky into a lot of trouble'. *Moi Otets*, p. 62. Homskaya refers to this 'notoriously famous telegram', Evgenia D. Homskaya, *Alexander Romanovich Luria: A Scientific Biography*, trans. by Daria Krotova (New York, NY: Plenum Press, 2001), p. 26. Eli Lamdan suggests the episode may be apocryphal. See Eli Lamdan, 'Who had Illusions? Alexander R. Luria's Central Asian Experiments on Optical Illusions', *Dubna Psychological Journal*, 3 (2013), 66–76, p. 71.

[28] Luria to Köhler, July 3, 1931, Philadelphia, American Philosophical Society (APS), Wolfgang Köhler Papers, Series I. Correspondence, Mss.B.K815.

[29] Luria, *Cognitive Development*, p. 100.

[30] Anne Harrington, *Reenchanted Science: Holism in German Culture from Wilhelm II to Hitler* (Princeton, NJ: Princeton University Press, 1996), p. 154.

Instead, Luria's writings tethered the capacity for abstract thought to Communist hopes for the Soviet Union.

According to Luria, abstract thinking allowed people to classify objects on the basis of an object's general features, which made it possible for people to think beyond the confines of their immediate experiences:

> The appearance of verbal and logical codes enabling one to abstract the essential features of objects and thus assign these objects to general categories … and permits conclusions to be drawn from given premises without having to resort to immediate graphic-functional experience.[31]

The transition from concrete to abstract thinking, which he also described as a transition from the sensory to the rational, is, he declared: 'a phenomenon that the classics of Marxism regarded as one of the most important in history'.[32] Typically, Luria's reference to the 'classics of Marxism' is unspecific. However, although he does not mention Lenin by name, Luria's conception of abstract thinking is consistent with Lenin's discussion of scientific abstraction outlined in his 'Conspectus on Hegel's Logic' and, as Sarah Marks has outlined, the 'predominant theory of mind in the Soviet Union, beyond Pavlovian studies of higher nervous activity, was Lenin's "theory of reflection"':[33]

> Thought proceeding from the concrete to the abstract … does not get away from the truth but comes closer to it. The abstraction of matter, of a law of nature, the abstraction of value, etc., in short all scientific (correct, serious, not absurd) abstractions reflect nature more deeply, truly and completely. From living perception to abstract thought, and from this to practice,—such is the dialectical path of cognition of truth, of cognition of objective reality.[34]

For Luria, as for Lenin, abstracting from nature does not drive a wedge between people and the world but allows them to understand it more fully.

[31] Luria, *Cognitive Development*, pp. 100–101.

[32] Luria, *Cognitive Development*, p. 101.

[33] Sarah Marks, 'Suggestion, Persuasion and Work: Psychotherapies in Communist Europe', *European Journal of Psychotherapy and Counselling*, 20, 1 (2018), 10–24, p. 22.

[34] V.I. Lenin, 'Conspectus on Hegel's Book The Science of Logic', *Collected Works*, vol. 38 (London: Lawrence and Wishart, 1961), pp. 85–244, p. 171. Vygotsky discusses this passage from Lenin's notes on Hegel in 'Development of Thinking and Formation of Concepts in the Adolescent' in *The Collected Works of L.S. Vygotsky*, vol. 5, ed. by Robert W. Rieber and trans. by Marie J. Hall (New York, NY: Plenum Press, 1998), pp. 29–82, p. 79. Luria discusses Lenin's theories of abstraction in *The Making of Mind*, pp. 177–178.

Luria's celebration of abstract thinking was combined with a scientific methodology that also operated with abstractions: his methodology relied on ordering, classifying and categorising. This chapter contends that by assigning Uzbek people to predetermined general categories, Luria's work did not always succeed in reflecting their nature more deeply, truly and completely in the manner he intended. In attending to the ambivalent qualities of Luria's research the argument of this chapter is similarly ambivalent, oscillating between sympathy with some of Luria's intentions and criticism of his methodologies and conclusions. It concludes by emphasising the tensions and strains evident on the pages of *Cognitive Development* between Luria and the people he conducted tests with, proposing that it might be generative to read the book against the intentions of its author. In contrast to Gielen and Jeshmaridian's unqualified condemnation, this approach is an attempt to trace the uncomfortable contradictions that characterised Luria's project rather than seeking to resolve them.

Revolution as Evolution

When outlining the goals of his experiments, Luria explicitly adopted a developmental framework compatible with the 'vanguard' temporality of the Communist Party. This unidirectional understanding of historical progress sat uneasily alongside his discussions of the temporal horizons he claimed were opened up by the transition from concrete to abstract modes of thought accompanying such progress, which emphasised sudden transformation and discontinuity. According to Luria, Uzbek people were not being passively carried up the steps of pre-existing universal time into the modern present. Instead, through education and social transformation, they would acquire the ability to actively create an as yet undetermined future, which might still open up the possibility of setting these very temporal structures into question. He declared (echoing Lenin again):

> Consciousness is the highest form of reflection of reality: it is, moreover, not given in advance, unchanging and passive, but shaped by activity and used by human beings to orient themselves to their environment, not only by adapting to conditions but in restructuring them.[35]

[35] Luria, *Cognitive Development*, p. 8.

In planning the Central Asian expeditions, Luria identified the simultaneous coexistence of groups at various 'stages' of development.[36] A loosely Marxist conception of history as advancing teleologically through a series of economic stages was combined with the Leninist conviction that such development could be artificially accelerated, conforming to what Francine Hirsch describes as 'state-sponsored evolutionism'.[37] Vygotsky wrote an essay in anticipation of the expeditions, which resounded with phrases snatched from the approved repertoire, and declared that:

> Instead of the colonial approach to the cultural development of backward peoples adopted ... in the capitalist world, this proposes completely new conditions for the development of a single socialist culture in different national forms.[38]

Luria similarly argued that collectivisation would propel supposedly 'backward' people swiftly forward, to make a 'leap of centuries'.[39] This was consistent with the Stalinist rhetoric that sought to make the USSR 'national in form, socialist in content' and also kept pace with the breathless tempo that characterised the era.

Following a strictly class-based analysis plucked directly from Marx's works was not possible in the Soviet Union, particularly in places like Central Asia without a sizeable urban industrial proletariat. This issue animated debates in the Kremlin and also impacted Luria's research. He chose to focus on five groups of people, all of whom were viewed as potential beneficiaries of Soviet power (categorised by the state as *bedniaks* or poor peasants), rather than those designated class enemies:

[36] The 'law of development by stages' [*po zakonu stadial'nosti*] was influenced by Engels's discussion of prehistoric societies. He claimed that human society progressed from savagery to barbarism to civilisation, a movement he compared to the evolutionary transition from ape to human. See Frederick Engels, *The Origin of the Family, Private Property and the State* (New York, NY: Pathfinder Press, 1972), pp. 47–54.

[37] Francine Hirsch, *Empire of Nations: Ethnographic Knowledge and the Making of the Soviet Union* (Ithaca, NY: Cornell University Press, 2005), p. 96. See Lenin, 'Imperialism, the Highest Stage of Capitalism', *Collected Works*, vol. 22, trans. by Yuri Sdobnikov (London: Lawrence and Wishart, 1964), pp. 185–304.

[38] Vygotsky, 'K voprosu o plane nauchno-issledovatel'skoi raboty po pedologii natsional'nykh men'shinstv' [On the Question of the Scientific Research Plan for the Pedology of the National Minorities], *Pedologiia*, 3 (1929), 367–377, p. 367.

[39] Luria, *Cognitive Development*, p. 164.

1. illiterate *ichkari* women in remote villages 'who were not involved in any modern social activities' (The term *ichkari*, meaning women's quarters, was used to refer to veiled women secluded within the home);
2. illiterate peasants, living in remote villages who were in no way involved with socialised labour and maintained the old economic way of life;
3. women who attended short-term courses in the teaching of preschool children (with no formal schooling and almost no training in literacy);
4. active *kolkhoz* (collective farm) workers and young people with considerable experience in planning production, distributing labour and taking stock of output but barely literate; and
5. women students admitted to teaching school after two or three years of study.[40]

Koffka's notes reveal that he interviewed an even broader range of participants, including miners and cooperative workers.[41] These categorisations took into account a range of interlocking factors including occupation, educational level, gender, age and place of residence. Koffka's notes also discussed the Uzbek counterparts that were sought for Russian terms: *bai* (or rich man) became a synonym for *kulak*, *dehqon* for *krest'ian* (peasant) and mullah for priest.[42] The bagginess of the terms used to categorise groups of people (both by Soviet officials and by Luria and his team of researchers) is another example of how the concrete continued to trouble the abstract.

Luria followed the Communist Party line, conceiving of his work in explicitly anti-imperialist terms and insisted that psychological propensities were not biologically determined.[43] Luria's rejection of biological

[40] Luria, *Cognitive Development*, p. 15. The same list appears in Luria to Köhler, July 3, 1931, APS, Wolfgang Köhler Papers, Series I, Correspondence, Mss.B.K815.

[41] Koffka's notes contain tables listing the age, ethnicity and jobs of the people he interviewed. See AHAP, Kurt Koffka Papers, Box M377, Folder 'Uzbekistan'.

[42] AHAP, Kurt Koffka Papers, Box M377, Folder 'Uzbekistan'.

[43] Historians disagree about the extent to which Soviet rule in Central Asia, contrary to party rhetoric, actually represented a continuity with the Tsarist Russian Empire. For an example of an account stressing continuities, see Edward Allworth, *Central Asia: 120 Years of Russian Rule* (Durham, NC: Duke University Press, 1989). For a counter example which stresses the novelty of the Soviet project, see Terry Martin, *The Affirmative Action Empire:*

essentialism was, however, coupled with a continued emphasis on cultural superiority framed in terms of historical development: a hierarchical framework that undermined his professed egalitarianism. Despite defining itself as anti-imperialist, his progressive framework for understanding Uzbek society bore comparison to the frameworks employed by Western anthropologists, whose work he drew on heavily.[44] Luria frequently drew parallels between Uzbekistan and other supposedly 'backward' places, including other regions of the USSR as well as communities in Africa.[45] For all his self-proclaimed attentiveness to cultural specificity, his emphasis on the interchangeability of 'backward' places understood cultural difference in terms of temporal development.

Johannes Fabian makes the sweeping claim that anthropology, borrowing from evolutionary theory (that was contemporaneous with its emergence as a discipline), naturalises time, removing it from history. For the anthropologist, relations between places are understood as relations between times; spatial dispersal is transformed into temporal sequence. Such schemas appear inclusive and universal but are founded on hierarchical separation. Anthropological time, true to its colonial origins, transforms horizontal difference in space into the vertical difference of time, justifying a logic of domination. Rendering space as time, according to Fabian, allows anthropologists to veil their qualitative judgements behind an apparently neutral, universal and natural abstract structure. Although there were specific ideas in the Soviet context about where 'backward' traits originated and how they might be eradicated, the projected dichotomy Fabian identifies between 'progress, development, modernity', on the one hand, and 'stagnation, underdevelopment, tradition' on the other, is evident in Luria's attitudes to Central Asia which

Nations and Nationalism in the Soviet Union, 1923–1939 (Ithaca, NY: Cornell University Press, 2001).

[44] See particularly, Luria and Vygotsky, 'Primitive Man and His Behaviour (1930)' in *Studies on the History of Behaviour: Ape, Primitive, and Child*, ed. and trans. by Victor I. Golod and Jane E. Know (Hillsdale, NJ: Lawrence Erlbaum, 1993), pp. 79–139. Luria attempted to invite the Austrian anthropologist Richard Thurnwald to accompany the expedition in 1932. See Luria to Köhler, July 3, 1931, APS, Wolfgang Köhler Papers, Series I. Correspondence, Mss.B.K815.

[45] Luria, *Cognitive Development*, p. 42 (on parallels with research into the perception of Zulu people). In a letter to Kurt Lewin Luria suggests that he is considering undertaking work in the Caucasus, Central Asia or Siberia, implying that the inhabitants of those places will have cognitive similarities. Luria to Lewin, July 12, 1929, AHAP, Kurt Lewin Papers, Box M2931, Folder 1.

similarly applied temporal frameworks to coeval societies.[46] In a similar vein, Gayatri Chakravorty Spivak's discussion of the 'violence of imperialist epistemic, social, and disciplinary inscription', which treats its objects of investigation as a deviation from an ideal, might also be applied to Luria's research (although in Luria's case the ideal subject was situated in the future).[47] Despite insisting that no way of seeing is 'a natural and inevitable achievement of the human mind', Luria's understanding of difference distributed humanity across one developmental slope and was thus implicitly value laden. For Luria, to borrow Yuri Slezkine's phrases: 'otherness was understood as nothing but backwardness,'[48] 'difference implied hierarchy'.[49] In *The Domestication of the Savage Mind*, Jack Goody praises Luria's work in Central Asia for drawing the conclusion that logic is based on 'highly artificial assumptions', but Goody overlooks Luria's assumption that even if there was not necessarily a more 'natural' solution to the problems posed in the experiments there was still a 'better' or more correct one.[50] Luria placed Uzbek people, culturally if not biologically, lower on the rungs of a single developmental ladder.

Luria and Koffka's personal accounts were pervaded by conflicting notions of time. Luria wrote that 'the masses had lived for centuries in economic stagnation and illiteracy',[51] suggesting that Soviet intervention would not merely speed up the progress of history but introduce history as such for the first time. Uzbeks were imagined to inhabit the past, but this past was conceived of as static, timeless. Upon arrival in Central Asia, Luria described an arid, monotonous and desolate landscape populated with

[46] See Johannes Fabian, *Time and the Other: How Anthropology Makes its Object* (New York, NY: Columbia University Press, 1983), p. 144.

[47] Gayatri Chakravorty Spivak, 'Can the Subaltern Speak?' in *Marxism and the Interpretation of Culture*, ed. by Cary Nelson and Lawrence Grossman (Houndsmills: Macmillan, 1988), pp. 271–313, p. 285. For a discussion of the application of postcolonial theory to the Soviet and post-Soviet context, see Gayatri Chakravorty Spivak, Nancy Condee, Harsha Ram and Vitaly Chernetsky, 'Are We Post-Colonial? Post-Soviet Space', *Modern Language Association*, 21, 3 (2006), 828–836 and David Chioni Moore, 'Is the Post-in Postcolonial the Post-in Post-Soviet? Toward a Global Postcolonial Critique', *PMLA*, 116, 1 (2001), 111–128.

[48] Yuri Slezkine, 'From Savages to Citizens: The Cultural Revolution in the Soviet Far North, 1928–1938', *Slavic Review*, 51, 1 (1992), 52–76, p. 57.

[49] Slezkine, *Arctic Mirrors*, p. 387.

[50] See Jack Goody, *The Domestication of the Savage Mind* (Cambridge: Cambridge University Press, 1977), p. 11.

[51] Alexander Luria, *Cultural Development*, p. vi.

Uzbek people with 'immovable faces' and 'motionless' eyes.[52] Psychologically, meanwhile, Uzbek people were said to exist in a perennially frozen now, incapable of conceptualising the future. In conformity with Edward Said's oft-cited analysis, Luria treats 'the Orient' itself as 'an ideal and unchanging abstraction' outside the dynamics of history.[53] As Frantz Fanon noted of the French attitude towards Algeria: the local people, 'the veiled women, the palm trees and the camels make up the landscape, the *natural* background' to the properly historical presence of the European.[54]

Orientalist tropes abound in Luria's personal impressions of Uzbekistan, all ancient 'whimsical' music, exotic smells and bustling bazaars which he compares to Baghdad.[55] Koffka's private letters indulge in similarly sentimental flourishes, adopting the idiom of the colonial adventure story, casting himself as the intrepid traveller boldly penetrating 'that vast expanse of night and scent, of softness and mystery', an exotic and eroticised feminine landscape lying supine before him.[56] He described one of his guides as 'quite medieval' and penned a poem in English for inclusion in the expedition's newspaper, which wistfully reflected on the inscrutability of the steppe.[57] Koffka also expressed disappointment that Uzbekistan did not consistently conform to his ideal romantic stereotype of the East, opining:

> Often in my youth had I dreamed of a journey to the East, of lying in the harbour of Port Said, the gate to all wonders of Asia. This was different. No proud ships, no glamorous Eastern port; a very commonplace car in a drab landscape of stunted firs and sand.[58]

[52] Luria, diary entry May 7, 1931 cited in E. Luria, *Moi Otets*, pp. 58–59.

[53] Edward Said, *Orientalism* (New York, NY: Random Books, 1978), p. 8. For a discussion of the applicability of Said's definition of Orientalism to the Russian Empire, see the debate between Adeeb Khalid, Nathaniel Knight and Maria Todorova in *Kritika*, 1, 4 (2000), 691–727.

[54] Frantz Fanon, *The Wretched of the Earth*, trans. by Constance Farrington (London: Penguin, 2001), p. 201.

[55] See the sections of Luria's diary from Central Asia reproduced in E. Luria, *Moi Otets*, pp. 59–60. In a particularly florid diary entry on 14 May 1931, Luria mentions both veiled women and camels, see E. Luria, *Moi Otets*, p. 60.

[56] Harrower, p. 158.

[57] Harrower, p. 150.

[58] Harrower, p. 157.

Although these experiments were conceived as part of a project to celebrate and encourage modernisation and rapid change, Koffka expressed dissatisfaction that Uzbekistan was too familiar, insufficiently trapped in an imaginary mystical past.

Luria's interest in Central Asia continued throughout his life. In 1964 he sent Jerome Bruner a postcard of a medieval mosque from a holiday in Samarkand saying he was enjoying the 'city of dreams'.[59] Luria frequently presented Western visitors with traditional gifts from Central Asia[60] and his Moscow apartment was decorated with handwoven Uzbek carpets.[61] Molly Harrower, an American psychologist and Koffka's erstwhile research assistant, recalled that on a visit to the USA in 1960 Luria insisted that she dress up in the traditional Uzbek outfit they had sent to her from the 1932 expedition.[62] Luria was more positive than Koffka about the changes that had occurred in Central Asia, however, noting in a postcard to Horace Kallen in 1932 that 'old Asian art and new forms of life' are combined in a 'wonderful mixture'.[63] In his travel notes he similarly observed with pleasure a conversation that lasted into the night combining 'the call to prayer, pillows and socialism'.[64]

These distinctions between Luria and Koffka's impressions are significant as Stalinist rhetoric in the years of the First Five Year Plan emphasised ethnic particularity rather than total homogeneity.[65] Stalin declared in 1929 that fostering distinct national identities was central to the goals of the revolution: 'on the ruins of the old, bourgeois nations new, socialist nations are arising and developing, and they are more solidly united than any bourgeois state.'[66] However, as Terry Martin discusses, the process of

[59] Luria to Bruner, May 16, 1964, Cambridge, MA, Harvard University Archives (HUA), Jerome Bruner Papers, HUG 4242.5 General Correspondence 1964–1965.

[60] For example, on 27 October 1967, Bruner wrote to Luria declaring: 'The camel has arrived!' (a gift Luria had sent from Central Asia). HUA Jerome Bruner Papers, HUG 4242.5 General Correspondence 1964–1665.

[61] E. Luria, *Moi Otets*, p. 57.

[62] Molly Harrower, 'A.R. Luria: A Personal Remembrance', AHAP, Molly Harrower Papers, Box 3212, Folder 23.

[63] Alexander Luria to Horace Kallen, August 12, 1932 (from Andijan, Uzbek SSR), AJA, Horace Kallen Papers, MS-1, Box 19, Folder 18.

[64] E. Luria, *Moi Otets*, p. 61.

[65] See Yuri Slezkine, 'The USSR as a Communal Apartment, or How a Socialist State Promoted Ethnic Particularism', *Slavic Review*, 53, 2 (1994), 414–452.

[66] J. Stalin, 'The National Question and Leninism (1929)', *Works*, vol. 11 (Moscow: Foreign Languages Publishing House, 1954), pp. 348–372, p. 356.

'indigenisation' [*korenizatsiia*] in practice involved the promotion of certain aspects of national culture alongside an attack on many fundamental practices: 'Promoting "national culture" meant aggressively promoting national identity, while undermining distinctive beliefs and social practices.'[67] The calculated promotion of national identity through folklore, national dress, cuisine, certain classic literary works and 'progressive' historic events resulted in a phenomenon Martin describes as 'Stalinist primordialism'.[68] Unlike in Orientalism as defined by Said, this involved the valorisation of approved symbolic markers of national identity alongside a programme of radical social transformation, including a campaign against religion. Hirsch similarly emphasises the differences between the left-wing folkloric impulses of Soviet reformers and their far-right German contemporaries. While German ethnographers focused on the primitive and eternal qualities of Germanic groups, in the Soviet Union traditional works were adapted to include distinctly socialist elements: Armenian folk songs about electric lighting, fairy tales about Lenin, Turkmen rugs embroidered with Stalin's face and folk dances that dramatised collectivisation.[69] The traditions of Central Asian republics may have been assumed to have no dynamic history of their own, but they were not understood as being eternally immutable now that history proper (socialism) had been introduced. Instead, tradition was framed as a dynamic aspect of a continually evolving historical society that would eventually be overcome. Luria's contradictory depiction of Central Asia, which simultaneously celebrated tradition and progress, was in keeping with this state discourse. This difference in attitude between Luria and Koffka—with the former emphasising change and the latter permanence—was also reflected in their differing experimental approaches and conclusions.

Luria Contra Gestalt (The Historical vs. the Universal)

The fundamental difference between the Gestalt approach and Luria's was the latter's insistence on the primacy of social factors in determining thought processes. Luria attacked Gestalt psychology for making universalising claims about human thought processes on the basis of experiments

[67] Martin, *The Affirmative Action Empire*, p. 183.
[68] Martin, *The Affirmative Action Empire*, p. 443.
[69] Hirsch, pp. 270–271.

conducted with people from a very limited demographic (primarily university educated and Western).[70] He argued that the Gestalt approach to psychology risked suffocating the richness of particularity under a homogeneous conceptual blanket.

In the preface to a Russian translation of Koffka, published shortly after the Central Asian expedition, Vygotsky clearly outlined his reservations with Koffka's work, claiming that in attempting to understand humans and apes using a single principle, Koffka neglected to probe what distinguished humans' thought processes from those of animals. Vygotsky deemed Gestalt theory too generalised, as it crucially 'ignores the historical nature of human consciousness'.[71] Koffka characterised development as natural and uniform, but, according to Vygotsky, human problems refuse to be straightjacketed in this manner, offering 'fierce resistance to the naturalistic attempt to interpret them [they] attempt to tear the cover of this single undifferentiated Gestalt to pieces'.[72] The differences between the two psychological approaches persisted in the conclusions that emerged from the experiments in Central Asia.

Both Luria and Koffka presented uneducated Uzbek people with various optical illusions intending to assess their perception of perspective. Luria concluded that respondents failed to respond to these tests, whereas Koffka asserted that responses were consistent with results obtained in the West; Luria emphasised cultural specificity and Koffka universality.[73] According to Koffka, the Uzbeks *did* have illusions; he attributed all exceptions to the 'suspicious' attitude of the participants, rather than to their cognitive abilities *per se*. For Koffka, thought structures did not differ in form between different cultures but only in content.

Luria's methodological approach ostensibly reflected his sensitivity to the cultural specificities of the people he was interviewing and differentiated his approach from Koffka's, who relied more heavily on standardised Western tests. Luria's intention was to make the experimental situation as 'natural as possible' by establishing an informal, 'unaggressive' and

[70] Luria, *Cognitive Development*, p. 31. Despite stating that his experimental approach was explicitly conceived in opposition to Gestalt psychology, Luria does not mention Koffka's participation in the expedition anywhere in *Cognitive Development*.

[71] Vygotsky, 'Preface to Koffka', *Collected Works of Lev Vygotsky*, vol. 3, ed. by Robert W. Rieber (New York, NY and London: Plenum, 1997), pp. 195–232, p. 216.

[72] Vygotsky, 'Preface to Koffka', p. 216.

[73] Compare Luria's conclusions in *Cognitive Development*, pp. 39–45 with Koffka's in Luria, 'The Second Expedition to Central Asia', p. 257.

'friendly' atmosphere.[74] Interviews were designed so that they would not interrupt habitual activities and mostly conducted 'in the relaxed atmosphere of a tea house … or in the camps in the fields and mountain pastures around the evening camp fire'.[75] Luria described how his team would begin by engaging people in a 'long, leisurely chat' before gradually introducing the questions that comprised the experimental protocol.[76] Conversations with women were conducted by women members of the research team in the home, where men were forbidden from entering. Luria was also alert to the limitations of using standardised tests developed in a Western context: 'Tests developed and validated in other cultures repeatedly produced experimental failures and invalidated our proposed study.'[77] The issue of standardised testing was prominent in Soviet psychological debates. For example, in 1930 Aron Zalkind rejected testing in Central Asia in characteristically bellicose fashion, declaring:

> [I]t is obvious that it is not that the children are imbeciles, but that the tests given to these children were imbecilic. Can we really ask questions devised on the basis of the socioeconomic and cultural experience of our own or Western capital cities? Can we really apply them to children of national minorities who live under completely unique conditions of the class struggle, of climate, culture, and everyday life?![78]

As such, the content of Luria's experiments was adapted to their Uzbek context.

Despite these efforts, however, Luria conceded that, 'no matter how natural we tried to make it and how much we prepared', the encounters were not always amiable and communication was not always easy.[79] All conversations were conducted in Uzbek through a translator and each psychologist was also accompanied by an assistant who sat away from the conversation in order to record people's responses as inconspicuously as possible. Not only was the conversation mediated, but Luria's commitment

[74] Luria, *Cognitive Development*, p. 16.

[75] Luria, *Cognitive Development*, p. 16.

[76] Luria, *Cognitive Development*, p. 46.

[77] Luria, *Cognitive Development*, p. 17.

[78] A.R. Zalkind, 'Psychoneurological Study of National Minorities', *Journal of Russian and East European Psychology*, 31, 1 (1993), 11–12 (from the stenographic record of a report presented at the plenum of the First All-Union Congress for the Study of Human Behaviour in 1930).

[79] Luria, *Cognitive Development*, p. 139.

to creating a 'natural' atmosphere wilfully overlooked the historical situation of the encounter where any translation from Uzbek into Russian implied a power relation. Diana Fuss makes a similar observation about Frantz Fanon's psychiatric work in Algeria: Fanon, who could speak neither Arabic nor Kabyle relied on translators, which, Fuss argues, 'could not avoid reproducing, within the space of the clinical treatment, the very structure of the colonial relation',[80] a problem Fanon later confronted, acknowledged and sought to overcome, as discussed in his co-written essay 'Social therapy in a ward of Muslim men: Methodological difficulties' (1954).[81]

In his diary, Luria noted with bewilderment that he and his team were often met with hostility and resentment by the people they were attempting to investigate. Luria observed that the mountain shepherds were particularly confused and resistant to questioning and claimed that Kirghiz women angrily implored the psychologists not to bewitch their children.[82] Anton Yasnitsky writes that

> Koffka noticed the extreme fear and anxiety in the group of the 'primitives' as their distinctive feature and suggested this was the reason for their differences ... with the group of the progressive socialist collective farmers and the like. The source of their fear and anxiety, in turn, may be explained by the fact that a military convoy of the notoriously brutal state security authorities participated in the expedition to the areas populated by the 'primitive' Uzbek shepherds in order to protect the members of the Soviet psychological expedition from local Islamist rebels, still very active in the area.[83]

Luria's experiments concluded that Uzbek people were reluctant to make suppositions about things beyond their immediate experience. One protocol involved presenting people with the following syllogism: 'In the Far North, where there is snow, all bears are white. Nova Zemlya is in the Far North and there is snow there. What colour are the bears there?' Many

[80] Diana Fuss, 'Frantz Fanon and the Politics of Identification', *Diacritics*, 24, 2/3 (1994), 19–42, p. 37.

[81] Frantz Fanon and Jacques Azoulay, 'Social Therapy in a Ward of Muslim Men: Methodological Difficulties' in *Alienation and Freedom*, ed. by Jean Khalfa and Robert J.C. Young, trans. by Steven Corcoran (London: Bloomsbury, 2018), pp. 353–372.

[82] This anecdote was recorded in Luria's travel notes on the 1932 expedition cited in E. Luria, *Moi Otets*, p. 61.

[83] Anton Yasnitsky, *Vygotsky: An Intellectual Biography* (London: Routledge, 2018), epub, p. 365.4.

respondents refused to infer anything from these words, stating bluntly: 'I don't know; I've seen a black bear' or 'We always speak of what we see; we don't talk about what we haven't seen.'[84] The hostility Luria encountered might be understood by pairing his own definition of Uzbek consciousness with a consideration of the historical situation within which the encounters took place: unlike white bears, Russian people were not unfamiliar to Uzbek people. Perhaps the hostile reactions of Uzbek people to the Russian psychologists had a basis in the historical experience of Russian imperialism and communism, rather than being the result of psychic 'backwardness', as Luria implied. In his analysis of Soviet policies in Central Asia, Douglas Taylor Northrop describes party documents of the period based on assumptions about Uzbek people that were similar to Luria's, but Northrop contends that

> Uzbek Muslims were not primitive or ignorant, as party writers thought: decades of tsarist rule in Central Asia had shaped quite rational opinions about who Russians were and what they wanted, and ten years of Bolshevik rule had added further notions about Communists, who mostly were seen as alien atheists and city folk.[85]

Uzbek people had their own assumptions and historically formed prejudices about Russian people, which influenced how they responded to the psychological tests.

In *Principles of Gestalt Psychology* (1935), a text that he began writing while convalescing from a fever in Central Asia,[86] Koffka imagined an encounter between a white person and a Chinese or Papuan person in native dress. Both the people, he wrote, will experience the other as strange. Although the other person seems to belong to the category 'person', the definition of which has been established by previous experience, he or she also departs from that definition in various ways. Koffka claims that if such encounters happen regularly then the category 'person' might be expanded to include people with different characteristics. He notes, however, that this resolution is unlikely. Instead, he argues, it is more frequently the case that

[84] Luria, *Cognitive Development*, p. 109.

[85] Douglas Taylor Northrop, *Veiled Empire: Gender and Power in Stalinist Central Asia* (Ithaca, NY: Cornell University Press, 2004), p. 165.

[86] Molly Harrower, 'A.R. Luria: A Personal Remembrance', AHAP, Molly Harrower Papers, Box 3212, Folder 23.

> the class schema remains unaffected and determines the characteristic of the individuals who, although they raise the claim to belong to it, deviate from it in certain striking aspects. The class schema, then, forms a sort of framework; what does not conform to the standard, appears as inferior. The stranger is the barbarian; he is inferior in every respect, simply because he is different from the type; he is less intelligent, less honest, less sensitive, and so forth.[87]

For Koffka, thought structures originate in experience but once established are resistant to change. He insisted that the schemas of thought between different groups of people are opaque to one another; they are identical in structure if not in content. This, however, assumed that the encounters between different groups of people are not influenced by historically produced structural inequalities as, for example, in a colonial situation. Instead, he described a situation of mutual and symmetrical incomprehension that failed to account for the domination of certain groups over others. Transposed into the Soviet context this observation does not account for the asymmetrical power relations between people from different class backgrounds, republics or ethnic groups.

Aesthetic judgement, Koffka continued, relies on a process of discrimination similar to that of the encounter between two foreigners; artworks are judged in relation to a pre-existing schema. He goes on to give an example from his experiences in Central Asia. In Samarkand, he reported, people were incapable of discerning the relative quality of an artwork. This inability to discriminate between works, he claimed, was due to the Islamic ban on images: Uzbek people were not historically exposed to pictures and thus developed no means of discriminating between them. This was disrupted by the revolution: 'with the old political order the religious order collapsed also, pictures were introduced and as a part of the new order were in themselves something good.'[88] According to Koffka, having had no education in forming aesthetic judgments for Uzbek people 'every picture is desirable … one sees the streets lined with photographers who have the most atrocious backgrounds, canvases painted in the most vulgar manner and representing ugly objects, against which the patrons stand to have their photographs taken'.[89] The language Koffka employs here is heavily value-laden—'atrocious', 'vulgar', 'ugly'—

[87] Kurt Koffka, *Principles of Gestalt Psychology* (New York, NY: Harcourt, Brace and Company, 1935), p. 349.

[88] Koffka, *Principles*, p. 350.

[89] Koffka, *Principles*, pp. 349–350.

redolent of his disdain for Uzbek culture. He assumed a lack of exposure to certain aesthetic principles resulted in inferior rather than just different tastes. Within the space of two paragraphs Koffka identifies how cultural prejudice operates before going on to betray his own negative assumptions about Uzbek perception in precisely the terms he had just outlined: as 'barbarian', 'inferior', 'less intelligent, less honest, less sensitive' than his own. Koffka acknowledged that people with different experiences operated using a different schema, but he did not entertain the possibility that competing ways of carving up reality are of equal validity, undermining his professed interest in the universal.

In *Distinction*, his sociological analysis of the relationship between taste and class in post-war France, Pierre Bourdieu discusses how aesthetic judgement, the process of differentiating between objects, participates in a system that consolidates differences between people. According to Bourdieu, power structures are reproduced through systems of classification: 'People's image of classification is a function of their position within it.'[90] In Bourdieu's account people both classify and are classified, but the process of classification is forgotten. Bourdieu provides a compelling account of how people's most mundane practices participate in maintaining existing social relations, positing taste as a crucial site of power: 'classificatory systems are … the stake of struggle between the groups they characterise and counterpose.'[91] But his argument is premised on an assumption of human psychological parity that Luria's research in Central Asia did not share. Bourdieu, like Koffka, assumes that classificatory systems rely on people who classify in the same manner but order things differently, whereas Luria set out to prove that the capacity to generalise *as such* was related to people's social position and cultural context.

Analysing how Uzbek people classified objects was one major strand of the psychological research conducted in Central Asia. For example, Luria conducted experiments with geometric shapes. When presented with a series of shapes, he found that people with some formal education would use abstract categories (circle, triangle, square etc.), whereas illiterate people would name them according to objects encountered in their everyday

[90] Pierre Bourdieu, *Distinction: A Social Critique of the Judgement of Taste*, trans. by Richard Nice (Cambridge, MA: Harvard University Press, 1984), p. 473. For a surprising account of a late Soviet engagement with Bourdieu's work that is itself an attempt to apply a Bourdieu-inspired methodology, see Georgi M. Derluguian, *Bourdieu's Secret Admirer in the Caucasus* (London: Verso, 2004).

[91] Bourdieu, p. 477.

lives (plates, beads, kettle stands etc.).[92] This would then influence how the subjects would group things together. Presented with three different sized rectangles, for example, one respondent designated them: a star, a door and a canal.[93] When asked whether the three shapes had anything in common, they responded that they did not as stars, doors and canals have nothing to do with one another in the real world. Illiterate people would only concede to group objects together by concrete situation: having identified two shapes as a glass and a bowl, one person replied that they could be placed together as they could both be used in a kitchen.[94] A similar pattern was found in experiments with colours: illiterate people would liken colours to objects they encountered in their daily lives (pig's dung, sky, pistachio, decayed teeth, cotton in blossom, etc.) and resisted the experimenters' suggestion that these might be grouped by brightness or saturation: 'they shook their heads in perplexity and failed to complete the task.'[95] In another experiment, when asked to exclude one object from a group of four, illiterate people grouped objects by situation rather than category: words 'were used not to codify objects into conceptual schemes but to establish the practical interrelations among things'.[96] Of course, this still describes a mode of classification, but for Luria this form of 'situational thinking' does not require that people make generalisations based on the 'essential feature of objects'.[97] Luria did not classify this mode of classification as classification because it was not 'advanced' (i.e. abstract) but understood as contingent on the vagaries of individual experience (i.e. not 'essential' but subjective). For Luria, literacy was the key factor which determined whether people organised objects conceptually rather than concretely; it was a prerequisite for 'advanced' forms of thought.

Language and Thought

Luria related some changes in thought to new collective forms of labour, claiming, for example, that collective farm workers were able to address 'pressing problems of social life', and arguing that the data gathered on

[92] Luria, *Cognitive Development*, pp. 34–39.

[93] Luria to Köhler, July 3, 1931, APS, Wolfgang Köhler Papers, Series I. Correspondence, Mss.B.K815.

[94] Luria, *Cognitive Development*, p. 37.

[95] Luria, *Cognitive Development*, p. 27.

[96] Luria, *The Making of Mind*, p. 73.

[97] Luria, *Cognitive Development*, p. 101.

the expedition 'adequately confirm that the mental life of these subjects changed radically because of collective social labour'.[98] Such statements, however, appear with relative infrequency in his conclusions. In practice, as Walter Ong observes, his conclusions mostly hinged on literacy, although he failed to 'systematically encode his findings expressly in terms of orality-literacy differences'.[99] Luria concluded that the more literate the person, the more abstract their thought processes. Indeed, Luria's assumptions about literacy seem consistent with Western European understandings dating back to the early modern period.[100] Luria's experiments set out to prove the hypothesis that literacy allowed people to think beyond their immediate material circumstances, implying that becoming literate had an intrinsic emancipatory potential.[101] Despite his persistent declarations to the contrary, Luria treated language as precisely the kind of universal structure that he identified and critiqued as a key limitation of the Gestalt approach to psychology, ignoring the very cultural and social specificities to which his work frequently claimed to be alert. But languages in the Soviet Union were volatile and, importantly, there were many different languages to choose between. Literacy was a highly charged political issue.

The Soviet Campaign to Eradicate Illiteracy (*Likbez*) was profoundly ideological, pivotal to the consolidation of Soviet power and the attempt to create new Soviet people with specific national identities. The Extraordinary Commission for the Liquidation of Illiteracy saw literacy as

[98] Luria, *Cognitive Development*, pp. 141–142.

[99] Walter J. Ong, *Orality and Literacy: The Technologizing of the Word* (New York, NY: Methuen, 1982), p. 50.

[100] 'Historically, literacy tends to be associated with the development of abstract thought, broadening of the mind through vicarious experience, the intensification and extensification of intellectual exchange, personal independence of thought and action, economic development. … Illiterates, on the other hand, are allegedly more restricted in their thought patterns, intellectually impoverished, culturally backward, isolated, inert, almost pathological.' R.A. Houston, *Literacy in Early Modern Europe: Culture and Education 1500–1800* (London: Longman, 1988), p. 2. See also, David Vincent, *The Rise of Mass Literacy: Reading and Writing in Modern Europe* (Cambridge: Polity, 2000).

[101] Luria's assumptions about literacy have been challenged by subsequent researchers who were directly inspired by his research in Central Asia. See James V. Wertsch, *Vygotsky and the Social Formation of Mind* (Cambridge, MA: Harvard University Press, 1985), p. 39. In his analysis of nineteenth-century England, E.P. Thompson discusses varying degrees of literacy, something Luria fails to consider. Thompson also insists that illiteracy 'by no means excluded men from political discourse' and considers the multifarious ways illiterate people actively engaged in political debates. E.P. Thompson, *The Making of the English Working Class* (London: Penguin, 1980), p. 782.

a key component in the building of socialism; a propagandistic mission was combined with a more practical belief that new forms of industry demanded literacy.[102] In the vast, multilingual Soviet Union establishing a set of codified national languages was a major element in the project of creating distinct national groups.[103] The Uzbek Soviet Socialist Republic (SSR) came into being in 1924 as part of the National Delimitation of the Central Asian Republics. The Soviet authorities, drawing on research by teams of mostly Russian ethnographers, imposed Sart (an Uzbek dialect from Samarkand) as the national language, standardising and codifying spelling, vocabulary and grammar to create new national norms. At the time of the revolution, Uzbek was written in various Arabic scripts, but in 1928 the Soviets replaced Arabic with the Latin alphabet, a policy related to the drive to eradicate Islam in the region.[104] Uzbek spelling was codified in 1929.[105]

Literacy levels reportedly soared from 3.8% in 1926 to 52.5% in 1932.[106] In practice, however, the linguistic situation in Uzbekistan remained heterogeneous and education chaotic, with most teachers brought in from Russia unable to speak the local languages.[107] Luria does not discuss how

[102] Although focussed on Russia, a useful overview is provided in Charles E. Clark, *Uprooting Otherness: The Literacy Campaign in NEP-Era Russia* (Selinsgrove: Susquehanna University Press, 2000). See also, Peter Kenez, *The Birth of the Propaganda State: Soviet Methods of Mass Mobilization, 1917–1929* (Cambridge: Cambridge University Press, 1985).

[103] See J.V. Stalin, 'Marxism and the National Question (1913)', *Works, Vol. 2, 1907–1913* (Moscow: Foreign Languages Publishing House, 1953), pp. 300–381 and J.V. Stalin, 'The National Question and Leninism', *Works, Vol. 11* (Moscow: Foreign Languages Publishing House, 1954), pp. 348–372.

[104] Luria mentions that 'removing the Muslim influence' was a major factor in the collectivisation campaign, *Cognitive Development*, p. vi. On anti-religious campaigns in the Soviet Union, see Daniel Peris, *Storming the Heavens: The Soviet League of the Militant Godless* (Ithaca, NY: Cornell University Press, 1998). On the Soviet 'assault' on Islam, see Shoshana Keller, *To Moscow, Not Mecca: The Soviet Campaign against Islam in Central Asia, 1917–1941* (Westport, CT: Praeger, 2001) and Adeeb Khalid, *Islam After Communism: Religion and Politics in Central Asia* (Berkeley, CA: University of California Press, 2007).

[105] On language policy in Soviet Central Asia, see Shirin Akiner, 'Uzbekistan: Republic of Many Tongues' in *Language Planning in the Soviet Union*, ed. by Michael Kirkwood (Basingstoke: Macmillan, 1989), pp. 100–121, Edward Allworth, *Uzbek Literary Politics* (The Hague: Mouton and Co., 1964) and Leonore A. Grenoble, *Language Policy in the Soviet Union* (Dordrecht: Kluwer Academic Publishing, 2003).

[106] Akiner, p. 118.

[107] See Michael G Smith, *Language and Power in the Creation of the USSR, 1917–1953* (Berlin and New York, NY: Mouton de Gruyter, 1998), pp. 121–142.

or why people were educated in any detail nor does he consider in which language his experimental subjects were being taught to become literate. Koffka's records of the 1932 expedition describe the extent to which language in Soviet Central Asia was far from a neutral, universal structure. His notes from a meeting with the Minister of Education from the Commissariat of Enlightenment of the People convey some of the disorganisation of the educational system in this period, with instruction in a mixture of languages (predominantly Russian).[108] He also expressed doubt about the statistics that he was quoted on literacy levels, as he met so few literate people.[109] Luria's discussion of literacy fails to account for the cognitive impact of this chaotic linguistic reality.

Furthermore, Koffka's private correspondence indicated that aside from Uzbek or Russian, Soviet people in Uzbekistan were also expected to become fluent in another language: the new language of Bolshevism.[110] Many loanwords entered Uzbek from Russian in the 1920s, which almost exclusively pertained to the new social structures and technologies introduced by the Soviet regime.[111] The content of reading materials available in state-run reading rooms and schools was also ideologically focused, including editions of Marx, Engels, Lenin and Stalin, newspapers and translations of approved authors. Instructional texts were produced with titles such as: 'What is a *Kolkhoz*?' 'What is a Soviet?' and 'The Party is Guiding Us'.[112] Koffka remarked upon the amazing homogeneity of opinion between the various educated people he encountered on his trip, noting their shared lexicon, historical narrative and ideological outlook:

> The uniformity of intellectual and emotional outlook is one of the strongest memories I carried away from my six weeks' visit to the Soviet Union. … To have built this wall in a relatively short time is one of the greatest achievements of the Soviet government.[113]

[108] AHAP, Kurt Koffka Papers, Box M379, 'Lectures Misc'.

[109] AHAP, Kurt Koffka Papers, Box M379, 'Lectures Misc'.

[110] On learning to 'speak Bolshevik', see Stephen Kotkin, *Magnetic Mountain: Stalinism as Civilization* (Berkeley, CA: University of California Press, 1997), pp. 198–237.

[111] Akiner, p. 105.

[112] This example comes from the Katanga national region in the Far North of Russia but is indicative of the kinds of materials being printed in non-Russian languages for non-literate and remote populations. Slezkine, *Arctic Mirrors*, p. 224.

[113] Harrower, p. 159.

Koffka was struck by the oppressive qualities of this tendency, but noted that these strangely homogeneous utterances were all the more surprising as they seemed to be made spontaneously and in earnest, noting that the people he encountered were not selected on purpose to impress or convert him.[114] On the night train to Tashkent, for example, he met a group of passengers from various Soviet Republics and backgrounds who 'toasted each other and the world revolution'.[115] Koffka's informal observations indicate how language and thought processes in the Soviet Union were becoming socialist in content, providing a new classificatory schema for evaluating the world.

Luria's cross-cultural experiments were preceded by a series of experiments undertaken by a group of psychologists investigating children in remote Siberia. Unlike the results of the Central Asian experiments, which went unpublished for four decades, detailed reports of the Siberian findings appeared contemporaneously, in an issue of the journal *Pedology* in 1929. These reports provide an insight into the pressing political concerns of the moment that are more muffled in Luria's later accounts, where he dispassionately presents socialism as a fait accompli. The psychologists conducting research in Siberia were concerned with analysing how educational policies impacted thinking. Many of their conclusions and methodologies similarly hinged on literacy and new forms of social life, but they also indicated the political urgency of conveying the meaning of particular concepts and institutions. One psychologist noted approvingly that 89% of children from the Tungus tribe recognised Lenin's portrait and 'correctly' declared that he 'did everything'.[116] This acquaintance with the face of the revolutionary leader, they noted, was the work of the cooperatives that acted like missionaries distributing his image: 'We found the portraits of leaders hanging in many yurts.'[117] This, however, only served to underline the distinction between visual and verbal forms of understanding: 'Knowing what a clock looks like (95%) was no guarantee that the Tungus

[114] Harrower, p. 159.

[115] Harrower, p. 159.

[116] I. Bulanov, 'Findings from a Study of the Tungus Child', *Journal of Russian and East European Psychology*, 31, 1 (1993), 51. These articles originally appeared in *Pedologiia*, 2 (1930).

[117] Bulanov, p. 52. Yuri Slezkine also describes ethnographers showing portraits of Lenin to tribes in the Russian Far North. One adult was reported to respond: 'Who's this bald-headed merchant?', *Arctic Mirrors*, p. 241.

knew our units of time.'[118] Similarly, recognising Lenin, it transpired, did not translate into an understanding of more abstract concepts like 'the Soviet Union' or 'Communism':[119]

> The children had a very vague notion of the USSR. Only a few even (20%) were familiar with the name, and of them only 5% (schoolchildren and students at the literacy centre) were able to say what 'USSR' meant. ... The Tungus children had no notion about the Revolution, the Communist Party, the Komsomol, the Pioneers, or the trade unions. Only 4% of schoolchildren knew what the Communists or the Pioneers or the Komsomol were. Most of the children had no notion of the Red Army. Only 7% (the schoolchildren) said that the Red Army was necessary.[120]

Another experiment concluded similarly that most children were 'still not fully aware that they had common interests with all the other workers in the Soviet Union'.[121] A psychologist who conducted experiments with adults was even more concerned: 'None of the adults could say what Moscow represented; they did not know, and could not explain, what Soviet power wanted, and they did not even know their regional centre, Verkhneudinsk.'[122] These people were not able to discuss places they had not personally visited, let alone define theoretical terms or newly introduced Soviet institutions.

Luria followed similar methods, asking people to define concepts like 'freedom'.[123] He reported that an illiterate 18-year-old woman from a

[118] Bulanov, p. 49.

[119] Luria draws a similar conclusion from a test involving the face of Lenin undertaken with young children in 1955. A two-year-old child was taught to turn its head to a portrait of Lenin and point to it when asked 'Where is Lenin?', but when the portrait was removed the child would continue to point to the same spot, indicating that they had not grasped the meaning of the word 'Lenin' but only the instruction to point. Luria, *Language and Cognition*, trans. by James Wertsch (Washington, DC: VH Winston and Sons, 1981), p. 47.

[120] Bulanov, pp. 51–52.

[121] K.I. Usova, 'The Tungus Child in School', *Journal of Russian and East European Psychology*, 31, 1 (1993), 37–44, p. 43.

[122] A. Shepovalova, 'The Everyday Social Environment of Tungus Children in the Northern Baikal Region', *Journal of Russian and East European Psychology*, 31, 1 (1993), 19–36, p. 35.

[123] This protocol bears comparison to a study undertaken on the language of Red Army soldiers in the mid-1920s, which asked similar questions and also concluded that recruits failed to assimilate the vocabulary of the Soviet regime (e.g., 75% of respondents did not understand terms like 'blockade', 'import', 'monopoly' and 'bureaucrat'). See Craig Brandist, 'Psychology, Linguistics and the Rise of Applied Social Science in the USSR: Isaak Shpil'rein's

remote village responded by saying: 'I've heard that women have got their freedom, but that's all I know. It means that landowners repressed them before but now they've escaped from their misery.'[124] This response may have indicated some rote-learned familiarity with official Soviet narratives but suggests that the word 'freedom' remained devoid of meaning, akin to the capacity to identify a clock without being able to tell the time.[125] (Though Luria did not consider that demands for freedom do not necessarily require a formal education in abstract concepts but might arise from concrete experiences of oppression, as C.L.R. James insists in *The Black Jacobins*: 'one does not need education or encouragement to cherish a dream of freedom'.) [126]

Luria was more approving of the response of a *kolkhoz* worker who had completed a ten-week agricultural course. When asked 'what is a cooperative?', the man asserted that shops were once run by merchants who sold goods to peasants at high prices but are now run by the government in the people's best interests and concluded by noting that cooperatives make people part of a community. Luria asserted that this person demonstrated a complex and independent understanding of emerging social forms: 'Defines in far more detail concept introduced by social system; makes greater use of abstract categories; clarifies one concept by means of another.'[127] For Luria, education was not understood as a method for pouring a particular body of knowledge into a passive receptacle but something that created a new kind of subject capable of actively engaging with complex ideas. However, it is unclear whether Luria would have been equally approving if the educated *kolkhoz* worker had responded by giving a detailed and eloquent critique of a cooperative. A denunciatory article published in 1934, which Luria protested mischaracterised his work, cited

Language of the Red Army Soldier' in *Politics and the Theory of Language in the USSR, 1917–1938*, ed. by Brandist and Katya Chown (London: Anthem Press, 2010), pp. 151–167, p. 161.

[124] Luria, *Cognitive Development*, p. 88.

[125] Luria notes that this woman, who, though illiterate, had attended some courses, was more advanced than those who had received no formal education. People with no formal education would define things tautologically ('A car is a car'), whereas she made some attempt to define things by description and comparison ('[A car] is smaller than a room, uses fire and people sit in it ... [like] cab drivers, bicycles, trains'), *Cognitive Development*, p. 88.

[126] C.L.R. James, *The Black Jacobins: Toussaint L'Ouverture and the San Domingo Revolution* (New York, NY: Vintage Books, 1963), p. 18.

[127] Luria, *Cognitive Development*, p. 89.

protocols from Luria's experiments not included in *Cognitive Development*, including a discussion with an illiterate *kolkhoz* chairman:

Question:	A rich landowner got the idea to cultivate cattle, and the cattle began to appear like flies.
The subject answers:	How is that possible that the cattle were like flies? He is a rich landowner, so his cattle are always as big as elephants.
Experimenter:	Assume that his cattle became like flies.
Answer:	We are moving toward socialism. The Soviet authorities took from the landowner all his possessions and all of his cattle and did not leave even a fly, i.e. cattle that were no bigger than a fly. The landowner's cattle cannot grow thin; if he only had cattle like flies, that would mean that the landowner had been dispossessed.[128]

As with the responses to the syllogism concerning white bears, Luria is said to have interpreted this as demonstrating a 'refusal to make assumptions', passing in silence over the subject's relation of the question to his first-hand experiences of collectivisation.[129] Here Luria's interest in tracing a specific form of individual development that he claimed was historically produced paradoxically led him to overlook an account of historical change. Luria asserted that abstract thinking allowed people to think outside the parameters of their everyday experiences and would accompany the kinds of social changes taking place in Central Asia, but his research conclusions suggest that he was less concerned with exploring how people responded to those changes than he was with encouraging them to develop the capacity to think in a way that was more remote from their everyday lives.

Revolutionary Assimilation

In 1932 the American anthropologist Melville J. Herskovits wrote to Luria to inquire about the Central Asian expeditions in the hope that they would shed light on his own research into African American communities in the USA. Herskovits described his work into 'New World Negroes' as

[128] Razmyslov, p. 52.
[129] Razmyslov, p. 52.

addressing 'how people from one civilization take over aspects of another'.[130] This resonated with arguments advanced in his 1928 book *The American Negro*, in which he wrote that the African American 'must solve the business of living in the culture that is not his own. How can he do it? He must learn to adjust himself to his cultural environment if he would survive.'[131] Herskovits characterises assimilation as a one-way process, in which the practices of one culture are simply replaced by those of the dominant culture without, however, eliminating social hierarchies based on racial prejudice.

The parallel Herskovits attempted to draw was not directly applicable to the Soviet context, however, where the social environment was still in flux and a Communist society was still in the process of being established. Indeed, the transformation in Central Asia formed part of a transformation expected of *all* Soviet citizens, Luria and his team of researchers included. Of course, Luria's research still implied that some were more equal than others, but unlike the relationship described by Herskovits, in which one oppressed culture was expected to disappear into the stable pre-existing culture of the oppressor, the question of the transformation required by Uzbek people was part of what Hirsch describes as a process of 'double assimilation': 'the assimilation of a diverse population into nationality categories and, simultaneously, the assimilation of those nationally categorized groups into the Soviet state and society.'[132] Furthermore, and unlike in American society as described by Herskovits, rather than referring to an already-established order of things, the ideal Soviet person was an abstract projection, anchored in the promise of a future yet to come.

Luria was quick to distinguish his work from Western studies which described the narrower world view of 'primitive' communities as an inherent deficiency of certain racial groups.[133] Despite drawing on their research, Luria differentiated his approach from Western anthropologists on the basis that their accounts of 'primitive' societies lacked any 'causal analysis':[134] they presented backwardness as an inherent and immutable

[130] Herskovits to Luria, March 2, 1932, NUA, Melville J. Herskovits Papers, General, 1906–1942, Box 13, Folder 4, 1931–1935.

[131] Melville J. Herskovits, *The American Negro: A Study in Racial Crossing* (New York, NY: Alfred A Knopf, 1928), p. 53.

[132] Hirsch, p. 14.

[133] Luria, *Cognitive Development*, p. vi.

[134] Luria to Kallen, October 29, 1931, AJA, Horace Kallen Papers, MS-1, Box 19, Folder 18.

aspect of the culture being studied rather than as something historically determined and therefore eradicable. Although Luria's research intentions were consistent with the broad campaign against biological determinism launched during the First Five Year Plan,[135] as he did not publish his findings immediately, the bitterly ironic outcome was that accusations of racism formed part of his denunciation. In 1935 he wrote to Herskovits to ask for advice on 'studies concerning the problem of racial differences in intelligence, and all what is [*sic*] published and sed [*sic*] by important men contra the fascistic ideas on racial inequality. We should be glad to show that such ideas are mere [*sic*] political and have no important scientific background.'[136] Written in uncharacteristically messy handwriting and poor English, this letter hints at the pressure Luria was under to demonstrate the invalidity of racial determinism. As part of his defensive response to criticisms, Luria explicitly contrasted his research with Nazi approaches in a 1933 article attacking German fascist science. He asserted that his psychological experiments proved the assumptions of German scientists wrong, but concluded by noting that investigations conducted by experts were unnecessary to demonstrate that thought processes are not biological: the revolution was proof enough. Liberating people from capitalism and 'colonial tsarist oppression' had a demonstrative impact on thought, testifying to the possibility for all people, regardless of their race or ethnicity, to participate in building the new society.[137]

Langston Hughes, who had moved to Harlem in New York shortly after Herskovits conducted research there, travelled around Soviet Central Asia in 1933. He also identified parallels between Central Asian people and African Americans, but unlike Herskovits he explicitly contrasted the Soviet Union and the USA.[138] Hughes spent 12 months in the Soviet Union in 1932 and 1933, arriving as part of a group of 22 African

[135] In the early 1930s, at the same time that Luria was conducting his experiments, Soviet ethnographers and anthropologists were sent to Central Asia with a mandate to disprove the theories of fascist 'race science' that was gaining increasing prominence in Germany. See Hirsch, pp. 246–272.

[136] See Luria to Herskovits, November 17, 1935, Northwestern University Archives, Melville Herskovits Papers, General, 1906–1942, Box 13, Folder 4, 1931–1935.

[137] Luriia, 'Psikhologiia ras i fashistskaia nauka' [The Psychology of Race and Fascist Science], *Front Nauka Tekhnikii*, 12 (1933), 97–108, p. 108.

[138] On the links between Herskovits's research and the Harlem Renaissance with which Hughes was associated, see: Mark Helbing, 'Feeling Universality and Thinking Particularistically: Alain Locke, Franz Boas, Melville Herskovits, and the Harlem Renaissance', *Prospects*, 19 (1994), 289–314.

Americans who were invited to perform in a film called ***Black and White***, an excoriating critique of racial segregation and the legacy of slavery in America slated to be directed by Luria's friend Sergei Eisenstein upon his return from Mexico.[139] The project was eventually abandoned, but the Soviet government invited the troupe to tour Soviet Central Asia. Members of the group expressed an interest in seeing what changes the October Revolution had wrought in the region, hoping to discover that communism was capable of eradicating racism, segregation and exploitation. Louise Thompson Patterson, a Communist Party USA member who organised the trip to the USSR, declared that she returned to America persuaded of this: 'What I had witnessed, especially in Central Asia, convinced me that only a new social order could remedy the American racial injustices I knew so well. I went to the Soviet Union with leftist leanings; I returned home a committed revolutionary.'[140]

Hughes pointed to parallels between the Jim Crow-era American South and the Tsarist 'South' in Central Asia: both were cotton-producing regions and both were home to racialised populations who laboured unfreely for their white masters. He characterised the distinction between pre- and post-revolutionary periods as a stark break not only in social life but in subjectivity: 'New times demand new people … in Soviet Asia, as elsewhere in the Soviet Union, new people are coming into being.'[141] His writings on his time in Central Asia consistently contrast the situation of non-white people in America and the Soviet Union. His travelogue *A Negro Looks at Soviet Central Asia* published in Moscow in 1934 opens with a two-page description of the horrors that define life in the American South—'segregation, peonage and lynching'. 'In America', he writes, 'if you are yellow, brown or black, you can never travel anywhere without

[139] See Steven S. Lee, 'Langston Hughes's "Moscow Movie": Reclaiming a Lost Minority Avant-Garde', *Comparative Literature*, 67, 2 (2015), 185–206.

[140] L. Patterson, unpublished memoirs, 'Chapter on Trip to Russia—1932', cited in Erik S. McDuffie, *Sojourning for Freedom: Black Women, American Communism and the Making of Black Left Feminism* (Durham, NC: Duke University Press, 2011), pp. 58–59. See also, Joy Gleason Carew, 'Black in the USSR: African Diasporan Pilgrims, Expatriates and Students in Russia, from the 1920s to the First Decade of the Twenty-First Century', *African and Black Diaspora: An International Journal*, 8, 2 (2015), 202–215.

[141] Langston Hughes, *A Negro Looks at Soviet Central Asia* (Moscow and Leningrad: Cooperative Publishing Society of Foreign Workers in the USSR, 1934), p. 49. See also, *Langston Hughes: Poems, Photos and Notebooks from Turkestan*, ed. by Zohra Said (New York, NY: CUNY Poetic Documents, 2015).

being reminded of your colour'.[142] The narrative of his Soviet journey then begins with him boarding an unsegregated train where he meets a man who he describes as being 'almost as brown as I am' and who he is amazed to discover is the mayor of Bukhara (noting that a black mayor in Mississippi would be unthinkable).[143]

After visiting a film school where illiterate nomads were being taught to operate film cameras, Hughes reflected on how 'impregnable Hollywood had been to Negroes', noting that black people were not even encouraged to operate cinema projectors, whereas in the Soviet Union he witnessed 'coloured people being taught by white men', a pedagogical process he perceived as empowering rather than paternalistic.[144] Hughes, unlike Koffka, shared Luria's political convictions, which meant that his romantic reveries about an imagined Orientalist Uzbekistan were similarly combined with a celebration of the modernisation ushered in by the revolution.[145] A more telling connection between Luria's and Hughes's respective writings on Soviet Uzbekistan, however, is the central position they both accorded to veiled and unveiling Muslim women.

Unveiling 'Primitive' Psychology

In Hughes's narratives of his time in Central Asia Uzbek women play a significant role. In both his fictional and autobiographical writings on the region, Hughes describes meeting with Uzbek women who have cast off their veils. In his 1934 travelogue, for example, he discusses meeting a woman named Halima Kasakova in Tashkent:

> with the Revolution still young in her part of the world, she took off the veil, went to school—and now? Well, now she is an important figure in the management of the Women's Club in the old part of the town, where for the first time in remembered history women sit on terraces open to the street and drink tea.[146]

Kate A. Baldwin argues that for Hughes unveiling was treated as synonymous with or at least as a metaphor for racial emancipation: 'The veil

[142] Langston Hughes, *A Negro Looks at Soviet Central Asia*, p. 6.

[143] Hughes, *A Negro Looks at Soviet Central Asia*, pp. 7–8.

[144] Langston Hughes, *I Wonder as I Wander: An Autobiographical Journey* (New York, NY: Thunder's Mouth Press, 1986), p. 116.

[145] Hughes, *A Negro Looks at Soviet Central Asia*, p. 28.

[146] Hughes, *A Negro Looks at Soviet Central Asia*, p. 49.

symbolised to Hughes not only the backwardness of gender segregation under Islam but also his own understanding of a predominant paradigm for racial consciousness in the United States as elaborated by [W.E.B.] Du Bois's *The Souls of Black Folks*', in which Du Bois claimed that a 'vast veil' separated black people from white people.[147] In Hughes's writings on Central Asia, Baldwin argues, unveiling functioned as a metaphor which envisaged liberation as a transgression of boundaries, combining two distinct discourses of the veil to create a new 'conceptual apparatus'.[148] For Hughes, the act of unveiling was understood not as a process of assimilation into an existing oppressive social order (as described by Herskovits), but as a revolutionary practice capable of breaking with or transgressing existing categories and hierarchies to create a new society.

Erik McDuffie and Baldwin both criticise members of the African American *Black and White* troupe who toured the region with translators employed by the state for their partial and distorted accounts of unveiling practices in Soviet Uzbekistan, which broadly agreed with the official discourses promoted by the Soviet authorities regarding the elimination of 'backwardness'.[149] McDuffie cites Chandra Talpade Mohanty in claiming that they viewed Soviet Central Asian women 'under Western eyes'.[150] Yet just as Said's definition of Orientalism does not fully apply to Soviet Central Asia, this critique is perhaps a little too neat, slotting this (mis) encounter into existing paradigms for understanding asymmetrical social relations in different contexts. While it is the case that Hughes's treatment of unveiling was primarily symbolic and based on a small number of interpersonal encounters with actual Uzbek women, Jennifer Wilson's suggestive reading of his 1934 article 'Boy Dancers of Uzbekistan', in which a celebration of the 'liberation' of Muslim women is combined with a mournful reflection on the fading out of the 'queer figure of the bachi' from Uzbek life, gestures towards a more complicated reading of his reflections on progress in the region that proposes a useful conceptual

[147] Baldwin, p. 102.

[148] Baldwin, p. 94.

[149] McDuffie is critical of Louise Patterson's impressions of unveiling in Soviet Central Asia, pp. 72–74.

[150] Chandar Talpade Mohanty et al., *Third World Women and the Politics of Feminism* (Bloomington, IN: University of Indiana Press, 1991), p. 74 cited in McDuffie, p. 72. An indication of the different valences of this phrase is to compare Mohanty's usage with Martin E. Malia, *Russia Under Western Eyes: From the Bronze Horseman to the Lenin Mausoleum* (Cambridge, MA: The Belknap Press of Harvard University Press, 2000).

distinction for approaching the dialogues with Muslim women included in Luria's *Cognitive Development.*[151]

Hughes's affirmation of revolutionary modernisation, Wilson argues, was combined with an affirmation of traditional queer practices, which undercut the progress narratives central to both Soviet revolutionary discourse (which framed homosexuality as a 'primitive vestige of the past'[152]) and black respectability politics in the USA. Wilson concludes that Hughes addressed 'an ethnic transnational awakened by Soviet internationalism' alongside a 'translocal queer collective [in Harlem and Tashkent] whose revolt was its own revolution'.[153] The tension in Hughes's work that Wilson identifies between two distinct revolutionary subjects—one identified with Soviet progress and modernisation, the other with traditional Uzbek practices within marginalised communities—is helpful to bear in mind when reading *Cognitive Development*, in which Luria's progressive frameworks for defining consciousness and cognition were often in tension with the utterances of the people he came into contact with. The modes of revolt, resistance and protest captured in the dialogues cited in *Cognitive Development* were incompatible with his definition of revolutionary consciousness but, following Wilson's reading of Hughes, could be interpreted as such nonetheless.

In his introduction to *Cognitive Development* Luria declares bombastically: 'When the socialist revolution eliminated dominance and submission as class relations, people oppressed one day enjoyed a free existence the next. And for the first time they enjoyed responsibility for their own future.'[154] In practice, however, his research conclusions do not bear this out. Not only was the pace of change slower than this statement suggests, but the groups Luria studied were not, as we have already seen, solely defined in terms of their class position. The study's focus on *ichkari* women was, however, consistent with Soviet policies in the region. In Central Asia, Soviet authorities targeted women as what Gregory Massell has called a 'surrogate proletariat', identifying Muslim women as an exceptionally oppressed group. Douglas Taylor Northrop even goes so far as to

[151] Jennifer Wilson, 'Queer Harlem, Queer Tashkent: Langston Hughes's "Boy Dancers of Uzbekistan"', *Slavic Review*, 76, 3 (2017), 637–646, p. 643.

[152] Wilson, p. 645. See also, Dan Healey, *Homosexual Desire in Revolutionary Russia: The Regulation of Sexual and Gender Dissent* (Chicago, IL: University of Chicago Press, 2001), p. 169.

[153] Wilson, p. 646.

[154] Luria, *Cognitive Development*, p. 13.

claim that in Uzbekistan 'women's emancipation ultimately came in many ways to exemplify the entire Bolshevik Revolution'.[155] In order to go about emancipating this group of revolutionary allies, an image of female inferiority within the Muslim society was constructed by mostly Western revolutionary elites.[156] In 1927 a *hujum* (assault or attack) was launched intending to liberate women from the oppressive traditions identified by Bolshevik ideologues, including family and kinship relations, property laws, religious practices, domestic roles and divisions of labour. In the face of a widespread violent backlash against unveiled women by Muslim men, the period during which Luria conducted his research and Hughes travelled the region saw a retreat from some of the more radical policies of the *hujum*.[157]

Koffka's notes from his introductory meeting with Uzbek Communist Party officials clearly foregrounded the issue of gender: 'Position of women: truly oppressed. Excluded from all culture. ... Now principle of equality, social and political.'[158] Luria wrote that '[t]he Islamic religion helped to maintain women's lack of rights', singling out their seclusion within the home and their practice of wearing a veil as evidence of oppression.[159] Muslim women were positioned between the modern Communist future and the traditional religious past with the *paranja*, a long robe traditionally made from horsehair, acting as the pre-eminent symbolic site

[155] Northrop, *Veiled Empire*, p. 9. For a vivid first-hand account of the social upheavals following the onset of the First Five Year Plan sympathetic to the Soviet cause, see Anna Louise Strong, *Red Star in Samarkand* (London: Coward-McCann, 1930).

[156] Gregory Massell, *The Surrogate Proletariat: Moslem Women and Revolutionary Strategies in Soviet Central Asia, 1919–1929* (Princeton, NJ: Princeton University Press, 1974), pp. 93–127.

[157] Edward A. Allworth, *The Modern Uzbeks: From Fourteenth Century to the Present* (Stanford, CA: Hoover Institute Press, 1990), pp. 223–224. The Women's Division ceased to operate in 1930. Unveiling may have symbolically ushered an Uzbek women into an abstract liberated collective, but in practice an isolated unveiled woman in an Uzbek village risked concrete violence. Thousands of women were murdered by local men, a widespread phenomenon that the Soviet administration did little to halt. See Shoshana Keller, 'Trapped between State and Society: Women's Liberation and Islam in Soviet Uzbekistan, 1926–1941', *Journal of Women's History*, 10, 1 (1998), 20–44. Though veil-wearing had generally decreased by the late 1930s, Marianne Kamp discusses the inconsistencies of veiling practices in Uzbekistan which differed from village to village, both because of differences in veil-wearing customs and because of differences in party policies on unveiling at a local level, pp. 218–220.

[158] AHAP, Kurt Koffka Papers, Box M379, 'Lectures Misc'.

[159] Luria, *Cognitive Development*, p. 14

in this ideological tussle. This was a contradictory position: uneducated Muslim women were simultaneously seen as the most socially repressed (and therefore the most revolutionary) but concomitantly the most psychically backward (and therefore the least revolutionary). Enter: the revolutionary vanguard.

In 'Algeria Unveiled' (1957), Frantz Fanon traces the ever-shifting status of veiling as a political strategy in colonial Algeria and during the struggle for independence. In the early 1930s, French colonial administrators in Algeria focused on the veil as a symbol of oppression in an analogous manner to their Soviet contemporaries. Fanon discusses how both veiling and unveiling went on to be adopted as political gestures of resistance at various moments in the revolution: Muslim women unveiled in order to disguise themselves as 'allies' of the Europeans, but at a later point in the conflict veils were reassumed in order to conceal weaponry. For Fanon, veiling in itself is neither inherently oppressive nor liberatory; what is paramount is the rejection of the coloniser: 'The phenomenon of resistance in the colonised must be related to an attitude of counter assimilation, of maintenance of a cultural, hence national, originality.'[160] All values imposed by the coloniser must be rejected 'even if these values objectively be worth choosing'.[161] As Adrienne Edgar argues, the Soviet Union's policies in Central Asia had more in common than with the modernising regimes of Ataturk's Turkey and Reza Shah's Iran than with, for example, those implemented in British India. Unlike in neighbouring Muslim countries, however, Soviet power in Uzbekistan was perceived as an alien, external force. Hence, according to Edgar, resistance to the policies was far more severe. This suggests, in line with Fanon's analysis, that the resistance to certain policies intended to 'liberate' Uzbek women was done so not on the basis of their content but at least in part because they were imposed by a foreign governing body.[162]

Throughout Luria's reports, illiterate *ichkari* women are shown to score the lowest on psychological tests as their 'immediate physical experi-

[160] Frantz Fanon, 'Algeria Unveiled' in *The New Left Reader*, ed. by Carl Oglesby (New York, NY: Grove Press, 1969), pp. 161–185, p. 167.

[161] Fanon, p. 184. Taylor Northrop's analysis agrees with Fanon's in that he argues that the veil became a major issue for Uzbek people as a result of its targeting as a symbol of oppression by the Bolsheviks.

[162] Adrienne Edgar, 'Bolshevism, Patriarchy, and the Nation: The Soviet "Emancipation" of Muslim Women in Pan-Islamic Perspective', *Slavic Review*, 65, 2 (2006), 252–272.

ence abounds'.[163] Luria characterises them as having a passive attitude. Yet it is these women who are described as the most intractable and resistant to the psychologists' suggestions. Transcriptions of interviews included in *Cognitive Development* reveal that the encounters were far from passive but sites of contestation, in which the women being interviewed obstinately refused to adopt the thought processes of their interlocutors. When asked to place two skeins of wool in different shades of orange together, for example, *ichkari* women were reported to reject the suggestion the skeins had anything in common: '"It can't be done", "None of them are the same, you can't put them together", "They're not at all alike" or "This is like calf's dung, and this is like a peach".'[164] Asked to respond to the syllogism concerning the existence of white bears in the Far North (cited above), the women interviewed 'refused to accept the major premise', 'replaced the inferential process by considerations of their own' or 'introduced general, rumour-based opinions about bears':

> "You would have to ask people who had been there and seen them." ... "There are different kinds of bears; if one was born red, he will stay that way." ... "The world is large, I don't know what kinds of bears there are." ... "Once I saw a bear in a museum, but that's all."[165]

Luria notes in a tone of exasperation that 'in each case they would avoid solving the task'.[166] In another exercise test subjects were asked to perform a basic mathematical problem: 'It takes thirty minutes to walk to village X, and it is five times faster on a bicycle. How long will it take on a bicycle?' Illi-Khodzk, a 24-year-old illiterate woman, responded: '"My brother in Dzhizak has a bicycle, and he goes much faster than a horse."' Nurmat, a 36-year- old illiterate woman replied: '"I know that bicycles go faster than bullock carts."'[167] Luria expresses frustration that the women related the question to their own experiences and that they refused to respond when the problem was repeated. Again and again Luria observes that despite prompting from the psychologists 'the subjects persisted in their own approach',[168] 'they refused decisively to draw inferences'[169] and that 'every

[163] Luria, *Cognitive Development*, p. 30.
[164] Luria, *Cognitive Development*, p. 27.
[165] Luria, *Cognitive Development*, p. 107, p. 111.
[166] Luria, *Cognitive Development*, p. 107.
[167] Luria, *Cognitive Development*, p. 121.
[168] Luria, *Cognitive Development*, p. 60
[169] Luria, *Cognitive Development*, p. 107.

attempt to suggest the possibility of categorical grouping met with protest'.[170]

Luria did not directly encounter these women, who were interviewed by his female colleagues in their homes. He only glimpsed *ichkari* women from afar, beneath their veils. Luria's account of the experiments, which agreed with Hughes's characterisations of the veil as being inherently oppressive, implied that these women's 'constrained' thought processes were a counterpart to their restricted social position, but the frustration that pervades his account of their responses to the experiments stems from their perceived inaccessibility, imperviousness to change and emphasis on their own lived experiences. In contrast, Fanon highlighted the limitations of his psychiatric frameworks rather than insisting on the psychic shortcomings of the Muslim women he conducted experiments with in Algeria, in which respondents similarly participated reluctantly and misunderstood the tasks he presented them with, concluding: 'Our patient's inadaptation is the correlate of the method's inadequacy'.[171]

Fanon describes the Western male frustration with the veiled Muslim woman as metonym for the colonial desire to conquer, understood in sexual terms (an eroticism which, as we have already seen, was evident in Koffka's writings on Uzbekistan). The veil implies an unknowability and distance that cannot be tolerated. The veiled woman sees without being seen and thus represents a threat to the colonising mission: 'There is no reciprocity. She does not yield herself, does not give herself, does not offer herself.'[172] In *The Wretched of the Earth*, Fanon also discusses the significance of non-cooperation within colonial regimes, identifying quotidian refusals to comply with the dominant order as acts of sabotage: 'The native's laziness is the conscious sabotage of the colonial machine; on the biological plane it is a remarkable system of auto-protection; and in any case it is a sure brake upon the seizure of the whole country by the occupying power.'[173] Fanon views such mundane acts of non-compliance as evi-

[170] Luria, *Cognitive Development*, p. 77.

[171] Fanon and Charles Geronimi, 'TAT in Muslim Women', *Alienation and Freedom*, pp. 427–432. Luria criticised projective tests for being insufficiently culturally specific in terms that agree with Fanon's analysis, but his approach nonetheless fails to connect Muslim women's responses to his experiments to their social experiences as Fanon does.

[172] Fanon, 'Algeria Unveiled' p. 169.

[173] Fanon, *The Wretched of the Earth*, pp. 237–238. This argument recalls Kotkin's discussions of the mundane interactions of people with the state that shaped the experience of Stalinism. He describes 'eager participation in, frequent circumvention of, and resourceful,

dence of a kind of proto-revolutionary subjectivity, the prelude to more militant, organised forms of struggle. His own psychiatric experiments with Muslim women emphasised the problem with tests in which 'no homogeneity exists between what the patient is presented with and what is known to her'.[174]

For Luria, however, *ichkari* women were understood as the antithesis of the ideal revolutionary subjectivity he hoped the social transformations accompanying collectivisation would bring into being. In his discussions of these women's responses, Luria proclaimed the forms of protest and resistance he encountered antithetical to revolutionary consciousness. Luria may have wanted to create active subjects capable of enthusiastically building a collective future, but that future was to be structured in a particular way and thus involved the imposition of a particular psychic framework. As such, the sparks of resistance in the encounters were not interpreted as flashes of spontaneity but as evidence of mental stultification blocking the path to consciousness.

This contradiction recalls the 'strife between normativity and the deconstruction of norms', which Fred Moten discusses in 'The Case of Blackness' (2008), an essay probing Fanon's work.[175] Moten discerns a fundamental ambivalence towards the pathological within Fanon's project, gesturing towards a tension between the normative impulse of the psychologist and the militant's desire to destroy all norms. Moten describes the contradictory position of the 'militant psychopathologist',[176] asking how it might be possible to reconcile adjustment, rehabilitation or reintegration with a politics that intends to radically transform the world: 'who would heal by way of explosion, excision, or exorcism'? [177] The question of the relationship between order and chaos, framed as a relationship between a social ideal and a subject perceived as psychically deficient, continues to animate the following chapters of this book which consider Luria's writings on children and people with aphasia and synaesthesia. Although Luria's work in Central Asia was not concerned with pathological cases *per*

albeit localised, resistance to the terms of daily life that developed within the crusade of building socialism', *Magnetic Mountain*, p. 21.

[174] Fanon and Geronimi, 'TAT in Muslim women', p. 431.

[175] Fred Moten, 'The Case of Blackness', *Criticism*, 50, 2 (2008), 177–218, p. 178. The tension between fact and lived experience is drawn from a mistranslation of a chapter title of Fanon's *Black Skin, White Masks*, which replaces the latter with the former.

[176] Moten, p. 208.

[177] Moten, p. 210.

se, insofar as Uzbek people were portrayed as primitive, backward and mentally disorganised, he did impute to uneducated Uzbeks what Moten describes as an 'ontology of disorder, an ontology of dehiscence',[178] defined in implicit opposition to the emerging Soviet 'order'. But ultimately the people he encountered 'ran away from the snares of recognition'.[179]

Moten writes 'that there is a counterpoint in Fanon, fugitive to Fanon's own self-regulative powers' and this chapter has proposed a similarly contrapuntal reading of *Cognitive Development*.[180] For Moten, an 'irremediable homelessness' is 'common to the colonized, the enslaved, and the enclosed'[181]; outside the coordinates of dominant structures that seek to contain them, these subjects threaten to unravel those structures through the refusal to be contained.

> The object vibrates against its frame like a resonator, and troubled air gets out. The air of the thing that escapes enframing is what I'm interested in—an often unattended movement that accompanies largely unthought positions and appositions.[182]

This troubled air wafts through the transcripts of Luria's experimental protocols. The subjects *Cognitive Development* describes vibrate against Luria's frames, but the misrecognitions, misunderstandings and disagreements documented in the book suggest the existence (if not the structure or content) of psychic lives beyond those frames. Indeed, the resistance Luria describes resides precisely in the *ichkari* women's refusal to adopt a form of thought that privileges framing (which relies on generalisation and abstraction) over scattered particularity. Rebecca Lemov's discussion of the case of a man named Tarev about whom the anthropologist Melford Spiro, a graduate student of Herskovits's, wrote in 'A Psychotic in the South Seas', whose 'failed' responses to a battery of tests exceeded the parameters of those instruments and so drew attention to their limitations, describes something similar to the excess evident in responses to Luria's protocols with *ichkari* women:

[178] Moten, p. 187.
[179] Moten, p. 211.
[180] Moten, p. 215.
[181] Moten, p. 187.
[182] Moten, p. 182.

> As a data source, he resisted capture, wittingly or unwittingly, yet indisputably. Tarev, with his exorbitant results and behavior, stood apart. … His answers to choreographed questions went beyond the scales designed to measure them.[183]

Luria claimed that the transition to socialism would be accompanied by a 'revolution' in thought that would eradicate rather than instal 'constricted' forms of thought, allowing people to break out from the confines of their everyday lives. In the context of these experiments, however, clinging to the immediate seems less constrictive than adopting the normative standards of the emergent dominant culture.

Conclusion

We have no way of knowing what became of the women Luria's collaborators interviewed, but an oral history project conducted with Uzbek women of the same generation in the 1990s found that those interviewed, most of whom ultimately became literate and unveiled, tended to present their lives as 'narrative[s] of progress' and reported that they identified many positive aspects of the late Soviet experience, which they contrasted to the hardships of collectivisation and the Second World War.[184] These descriptions of 'progress', however, seem to have been grounded in material changes in living standards and working conditions, rather than indicating the adoption of an abstract teleological framework for understanding historical development.

Before leaving Moscow for the second expedition to Central Asia in 1932, Luria, Koffka and Vygotsky attended a screening of Eisenstein's *The General Line* (1929), which was released in Russian under the approved title *The Old and the New*. The film depicts the transformative impact of technology on the traditional agrarian countryside. In her discussion of the film's relation to contemporaneous Soviet discourse, Anne Nesbet cites a bellicose speech delivered by Stalin in 1927:

> That which is dying out does not just want to get on with it and die, but rather fights for its existence, insists on its own outlived business. Here there

[183] Rebecca Lemov, 'On Being Psychotic in the South Seas, Circa 1947', *History of the Human Sciences*, 31, 5 (2018), 80–105, p. 94.

[184] Marianne Kamp, *The New Woman in Uzbekistan: Islam, Modernity, and Unveiling under Communism* (Seattle, WA: University of Washington Press, 2006), p. 228.

> is always something being born into life something new. But that which is born is not simply born, but wails and cries, insisting on its right to exist. (Voices: 'True!' Applause). The battle between old and new, between that which is dying out and that which is being born—this is the basis of our development.[185]

Stalin depicted the fight between old and new ways of life, tradition and modernity, capitalism and communism, by imagining the latter as a vulnerable newly born baby. Luria's experiments in Central Asia celebrated the birth of the new but bore witness to the tenacity of the old (and it was for this reason that his work was eventually denounced). This struggle between the old and the newly born also informed Luria's work on child psychology. The study of non-metaphorical newly born babies was central to the question of historical development posed by Stalin. However, as we shall see, the wails and cries of those born into Soviet life were sometimes subordinated to the voices of those conducting their 'own outlived business': adults.

[185] Joseph Stalin, XV s"ezd vsesoiuznoi kommunisticheskoi partii (b), [opened December 2, 1927] (Moscow and Leningrad: Gosudarstvennoe izdatel'stvo, 1928), p. 71, cited in Anne Nesbet, *Savage Junctures: Sergei Eisenstein and the Shape of Thinking* (London: IB Tauris, 2003), p. 98.

CHAPTER 4

The Child

In 1958, shortly after the onset of the 'thaw' when Luria began travelling to places outside the Soviet Union again, he met the British psychoanalyst D.W. Winnicott at an international conference in Copenhagen. Winnicott subsequently contacted Luria to ask whether his paper 'The First Year of Life' would be of interest to Russian psychologists, concluding his letter by noting: 'It interests me very much that babies are really the same everywhere.'[1] Written at the height of the Cold War Winnicott's comment seems intended as an oblique gesture of empathy across the ideological divide.[2] Regardless of his personal views, the continued restrictions on psychoanalytic publications in the Soviet Union would presumably have prevented Luria from facilitating a translation. Despite Luria's early engagements with psychoanalysis, the conceptualisation of the child he began to elaborate in the 1920s was markedly different in emphasis from psychoanalytic accounts developed in post-war Britain, and his exposure to ongoing psychoanalytic debates after 1930 was limited. Like Winnicott,

[1] D.W. Winnicott to Luria, July 7, 1960, London, Wellcome Collection (WC), D.W. Winnicott Papers, PP/DWW/B/B/8, Letters 1938–1968. This is the only letter to Luria in the Winnicott archive suggesting that Luria did not respond to his request.

[2] Winnicott went on to write two essays in which the Berlin wall figured as a metaphor for psychic life: D.W. Winnicott, 'The Value of Depression' (1963) and 'Berlin Walls' (1969), in *Home is Where We Start From: Essays by a Psychoanalyst*, ed. by Claire Winnicott, Ray Shepherd and Madeleine Davis (New York, NY: W.W. Norton, 1986), pp. 71–79 and pp. 221–227.

H. Proctor, *Psychologies in Revolution*, Mental Health in Historical Perspective, https://doi.org/10.1007/978-3-030-35028-4_4

Luria was concerned with the child's adjustment to its environment, the development of the imagination and language acquisition, but the child's experiences in the family, gender difference and childhood sexuality—some of the central tenets of psychoanalytic theories of childhood—are conspicuous by their almost complete absence from Luria's writings on child psychology. As Denise Riley discusses in her analysis of post-war British developmental psychology *War in the Nursery* (1983), 'generally held notions—of what an individual is, what society is—slip, uncriticised into child psychology'.[3] The absence of certain concerns from Luria's work signals the presence of an understanding of individual development organised according to different precepts and elaborated in a specific social and political milieu where notions of what individuals and society were or might become were still unsettled.

Unlike Vygotsky, who is renowned as an educationalist and developmental psychologist in the West, Luria's Anglophone reputation is primarily tied to his contributions to neurology, but Luria also published works on child psychology throughout his long career.[4] This chapter focuses on a cluster of texts based on experiments conducted by Luria and his collaborators between 1923 and 1936, though some were only published many years later. It begins by considering Luria's more general theoretical discussions of early infancy, play and language acquisition from this period, including works written in collaboration with Lev Vygotsky and some

[3] Denise Riley, *War in the Nursery: Theories of the Child and Mother* (London: Virago, 1983), pp. 7–8.

[4] Evgenia D. Homskaya, *Alexander Romanovich Luria: A Scientific Biography*, trans. by Daria Krotova (New York, NY: Plenum Press, 2001), pp. 127–161. An international bibliographical overview on Vygotsky demonstrates this tendency in the secondary literature: Mohamed Elhammouni, *Socio-Historicocultural Psychology: Lev Semenovich Vygotsky (1896–1934), Bibliographical Notes* (Lanham: University Press of America, 1997). For an example of the application of Vygotsky's work aimed at practicing educators, see Robert Lake, *Vygotsky on Education: A Primer* (New York, NY: Peter Lang, 2012). A critical appraisal of the influence of Vygotsky on Western Developmental Psychology is given in Erica Burman, *Deconstructing Developmental Psychology* (Hove: Routledge, 2008), pp. 241–260. This chapter does not engage with the large international body of work engaging with and extending Vygotsky's and Luria's research in the field of child psychology by practicing psychologists and educators published from the 1950s onwards, nor does it attempt to give an account of the concepts that became most famously associated with their 'cultural-historical psychology', for example, the 'zone of proximal development'.

texts written solely by Vygotsky.[5] The chapter then considers various publications based on experiments conducted with identical and fraternal twins in a laboratory setting at the Medico-Genetic Institute in Moscow between 1934 and 1936, before concluding with a discussion of Luria's 'A Child's Speech Responses and the Social Environment' from *Speech and Intellect among Rural, Urban and Homeless Children* (1930), a work comparing the cognitive capacities of children from different social backgrounds.[6]

This relatively short interval of time saw Luria work in a range of institutions and includes the period he spent in Kharkiv. Alongside working at the Institute of Psychology in Moscow, Luria (with Vygotsky and a team of collaborators including Lidia Bozhovich, Roza Levina, Natalia Morozova, Lia Slavina and Aleksandr Zaporozhets) worked at the Academy of Communist Upbringing named after Nadezhda Krupskaya, where he headed the psychological laboratory.[7] Luria later reflected that he, who had no Party affiliation at that time and was still very young, felt out of place among the fervently ideological 'grown-up Party activists' who ran the place, though he nonetheless contended that despite the apparently more zealous commitment of others at the institution to the revolutionary cause, Vygotsky's approach was the only 'real, not sham'

[5] Although Luria and Vygotsky analysed young people at various stages of development, from birth to adolescence, my focus in this chapter will primarily be on children under seven, typically referred to as 'preschoolers' [*doshkolniki*] in the literature.

[6] Luria also engaged with the discipline of 'Defectology' [*defektologiia*], a discipline dedicated to studying and caring for disabled children. A discussion of his publications and research in this field is beyond the scope of this book. Luria worked at the Institute of Defectology of the Academy of Pedagogical Sciences between 1953 and 1959. See *Umstvenno otsalyi rebenok: Ocherki izucheniya osobennostei vysshei nervnoi deytel'nosti detei-oligofrenov* (Moscow: Academy of Pedagogical Sciences of the RSFSR, 1960) translated as A.R. Luria ed., *The Mentally Retarded Child: Essays based on a Study of the Higher Nervous Functioning of Child-Oligophrenics*, trans. by W.P. Robinson (London: Pergamon Press, 1963). On defectology after the 1936 decree banning pedology, see Andy Byford, 'The Imperfect Child in Early Twentieth-Century Russia', *History of Education*, 46, 5 (2017), 595–617. For an insight into Luria's continued engagement with educational institutions, see the British educationalist Brian Simon's accounts of a trip to the USSR in 1961. Luria acted as Simon's host coordinating a trip to psychological and educational institutes in Moscow, Leningrad and Tbilisi. See London, Institute of Education (IOE), Brian Simon Archive, Book 5 1957–1963, DC/SIM/1/72.

[7] See T.V. Akhutina, 'A.R. Luriia: zhiznennyi put'' [A.R. Luria: The Way of Life], *Kul'turno-istoricheskaia psikhologiia*, 2 (2012), 2–10, p. 3 and Anton Yasnitsky, *Vygotsky: An Intellectual Biography* (London: Routledge, 2018), epub, pp. 240.3–244.6.

Marxist psychology developed in the period.[8] In 1930 the Academy of Communist Upbringing's School of Social Sciences was proclaimed 'Trotskyist' and closed, relocating to Leningrad in 1931.[9] Vygotsky went to work (elsewhere) in Leningrad while Luria and a group of collaborators moved to Kharkiv, which was then capital of the Ukrainian SSR. Luria returned to Moscow entering the Medico-Genetic Institute in October 1934 as director of the laboratory of psychopathology where he undertook experiments with twins. Vygotsky had died in June of that year.

Though Vygotsky was more closely associated with the discipline than Luria, their work in this period coincided with the brief ascendance of pedology in the Soviet Union. Pedology was not born on Soviet soil. Its first practitioner was G. Stanley Hall, the American psychologist responsible for organising Freud's visit to the USA in 1909. A pedological institute was founded in Russia that year. As the discipline dwindled in importance in Western Europe and America, its popularity grew in the Soviet Union, reaching its peak as Stalin was consolidating his power. Pedology was intended as the comprehensive science of the child, yet so divergent are secondary accounts of the discipline's history that Alexei Kozulin's assertion that, 'almost all work in educational psychology in the 1920s was called "pedology"', seems to accurately summarise the scope of the project.[10] In April 1928, the Planning Commission for Pedological

[8] See K.E. Levitin, 'A Criminal Investigation', *Journal of Russian and East European Psychology*, 36, 5 (1998), 46–75, pp. 74–75.

[9] A.A. Leontiev, 'The Life and Creative Path of A.N. Leontiev', *Journal of Russian and East European Psychology*, 43, 3 (2005), 8–69, p. 31.

[10] Alexei Kozulin, *Psychology in Utopia: Toward a Social History of Soviet Psychology* (Cambridge, MA: MIT, 1984), p. 22. For accounts of the history of pedology in the Soviet Union, see E. Balashov, *Pedologiia v Rossii v pervoi tret'i XX veka* [*Pedology in Russia in the First Third of the 20th Century*] (St Petersburg: Nestor-Istoriia, 2012), Raymond Bauer, *The New Man in Soviet Psychology* (Cambridge, MA: Harvard University Press, 1952), pp. 116–127, *Making Education Soviet, 1917–1953*, ed. by Andy Byford and Polly Jones (London: Routledge, 2006), Alexander Etkind, 'Pedological Perversions' in *Eros of the Impossible: The History of Psychoanalysis in Russia*, trans. by Noah and Maria Rubins (Boulder, CO: West View Press, 1997), pp. 259–285, David Joravsky, *Russian Psychology* (Oxford: Blackwell, 1989), pp. 345–354, Elena Minkova, 'Pedology as a Complex Science Devoted to the Study of Children in Russia: The History of its Origin and Elimination', *Psychological Thought*, 5, 2 (2012), 83–98 and A.V. Petrovskii and M.G. Iaroshevskii, *Psikhologiia* (Moscow: Akademia, 2002), pp. 146–147. For a useful anthology of primary texts in translation, see F.A. Fradkin ed., *A Search in Pedagogics*, trans. by Peter Emerson (Moscow: Progress Publishers, 1990). An early bibliographical overview is provided in N. Rybnikov, *Russkaia Pedologicheskaia Literatura* (Orel: Krasnaia Kniga, 1925).

Research in Russia was set up. Former psychoanalyst turned fervent Communist ideologue Aron Zalkind was appointed its chairman. The journal *Pedology* also started that year. The first All-Union Pedological Congress was held in 1928 and ran for a week. Approximately 3000 delegates attended the event, including high-ranking party leaders. Nikolai Bukharin, leader of the 'right opposition', addressed the Congress in a speech that lasted two hours, laying the foundations for the new science's methodological approach. The link between children and the industrial future of the Soviet state was clearly stated:

> We need to direct our strength not into abstract chatter, but into an effort to produce a certain number of living workers in the shortest time frame; qualified, specially schooled machines that we can start up right away and set into motion.[11]

This statement by the practicing pedologist M. Gel'mont conveys the rhetoric typical of the discipline:

> The foundations of the old social order are disappearing under the huge building of socialism. A new socialist life is being created and a human transformation is occurring. The old proprietary and individualistic habits are being eradicated and a communistic attitude towards work and life is being formed.[12]

Pedologists were to play a crucial role in education. Most schools across the Soviet Union had a resident pedologist and teachers were encouraged to record the behaviour of their students.[13] By 1930 the Institute of Psychology in Moscow was renamed the Institute of Psychology, Pedology

[11] From Speeches at the First Pedological Conference (1928), cited in Etkind, *Eros of the Impossible*, p. 265.

[12] M. Gel'mont, 'Pedologo-Pedagogichekoe Izuchenie Kollektivizirovannogo truda i byta' [Pedological and Pedagogical Research in the Field of Collective Work and Everyday Life], *Pedologiia*, 1, 13 (1931), 17–23 (citation a modification of the summary on p. 71). Aron Zalkind gives a characteristically forceful programmatic outline of pedology in *Pedologiia v SSSR* [*Pedology in the USSR*] (Moscow: Rabotnik Prosveshcheniia, 1929).

[13] On the impact of pedology on Soviet schools, see Larry E. Holmes, *Stalin's School: Moscow's Model School no. 25, 1931–1937* (Pittsburgh: University of Pittsburgh Press, 1999), pp. 137–141.

and Industrial Psychology, but in 1936 pedology was banned, which, as we shall see, had a decisive impact on Luria's subsequent career trajectory.[14]

Children of the Revolution

The figure of the child was central to Soviet ideology. Were babies all born alike? What influence did hereditary, class, nationality and education have on the child? What was the optimal upbringing for a Soviet person? What would happen to the family under communism? What role would the state play in rearing children? The two decades following the revolution witnessed profound shifts in the Soviet state's attitude to children, yet as Lisa Kirschenbaum argues, 'the happy child as icon of socialist transformation remained'.[15] In the aftermath of the revolution, bold utopians dreamt of the withering away of the family and the construction of a new collective existence:

> The new cities will bring about such drastic changes in life relationships that the words 'my children' and 'my parents' will lose the meaning of ownership and will no longer be used. An entirely new scheme of life will develop when a family of several thousand people live in a commune, with all the children growing up together and every adult contributing his share for their education.[16]

The institutions and relationships that formed the basis for a child's upbringing were all subject to change: gender roles in the workplace, family structures, housing, dining, education and childcare. Often contradictory ideas aimed at eviscerating old forms of everyday life [*byt*] challenged almost every aspect of quotidian existence. The 1918 family code prohibited children from being adopted by individual families, positing the state

[14] See 'On Pedological Distortions in the Commissariats of Education. A Resolution of the Central Committee of the Communist Party of the Soviet Union July 4, 1936', Joseph Wortis, *Soviet Psychiatry* (Baltimore, MD: Williams & Wilkins, 1950), 242–245. For the original, see: 'O pedologicheskikh izvrashcheniiakh v sisteme narkomprosov (4 iiulia 1936 goda)' in *KPSS v rezoliutsiiakh i resheniiakh s"ezdov, konferentsii i plenumov TsK*, vol. 6 (Moscow: Politizdat, 1985), pp. 364–367. See also: N. Kurek, *Istoriia likvidatsii pedologii i psikhotekhniki* [*The History of the Liquidation of Pedology and Psychotechnics*] (St Petersburg: Aleteia, 2004).

[15] Lisa A. Kirschenbaum, *Small Comrades: Revolutionizing Childhood in Soviet Russia, 1917–1932* (London: Routledge Falmer, 2001), p. 163.

[16] William Clark Trow, *Character Education in Soviet Russia* (Ann Arbor, MI: Ann Arbor Press, 1934), p. 28.

as the guardian of its young citizens in an attempt to radically reimagine traditional family structures. The fledgling Soviet government legalised divorce and abortion, secularised marriage and encouraged women to enter the workforce in unprecedented numbers.[17] However, women themselves, who were often depicted as neglectful or incompetent mothers in need of state guidance, were not always seen as the prime beneficiaries of these policies. According to Lynn Mally, 'women embodied all the problems of the past, but children were the hope of the future'.[18] New baby names proliferated in the years after the revolution, accompanied by new atheistic naming ceremonies. Populating the world with 'humans called Spark, Joy, Will, Electric, Rebel and Barricade' or variations on Marx, Engels, Robespierre and Rosa Luxembourg was an aspirational gesture, as if the very act of naming alone might bring an emancipated world into being.[19]

Communism, however, could not be constructed overnight. As with everything in the post-revolutionary period, the concrete conditions of the present chafed uncomfortably against abstract visions of the future. Despite the radical concrete changes that were wrought in Soviet social life, the impact and implementation of reforms were often hampered by material constraints and more traditional structures and practices persisted. Soviet theories of childhood, like actual Soviet children, were forced to contend with the realities of a country ravaged first by war and revolution and later by rapid industrialisation and the collectivisation of agriculture. By the time the Bolsheviks seized power, maintaining the *status quo* was not even an option, but a new social order structured according to a new set of norms gradually began to emerge from the wreckage of the old. As the revolution faded on the historical horizon, a stable world began to be built, which still paradoxically defined itself as being characterised by

[17] On the impact of Soviet legislation and Communist Party rhetoric on women, see Wendy Z. Goldman, *Women, the State, and Revolution: Soviet Family Policy and Social Life, 1917–1936* (Cambridge: Cambridge University Press, 1993), Richard Stites, *The Women's Liberation Movement in Russia: Feminism, Nihilism, and Bolshevism, 1860–1930* (Princeton, NJ: Princeton University Press, 1978) and Elizabeth A. Wood, *The Baba and the Comrade: Gender and Politics in Revolutionary Russia* (Bloomington: Indiana University Press, 1997).

[18] Lynn Mally, *Culture of the Future: The Proletkult Movement in Revolutionary Russia* (Berkeley, CA: University of California Press, 1990), p. 180. On paternalistic propaganda aimed at mothers, see David Hoffman, *Stalinist Values: The Cultural Norms of Soviet Modernity, 1917–1941* (Ithaca, NY: Cornell University Press, 2003), pp. 88–117.

[19] Richard Stites, *Revolutionary Dreams: Utopian Vision and Experimental Life in the Russian Revolution* (Oxford: Oxford University Press, 1989), pp. 111–112.

immolation and innovation. Situated between revolutionary transformation and the consolidation of an emergent order of things, the contradictory figure of the revolutionary child embodied the paradox of Soviet society, caught between routine and rupture, revolution and reproduction, transformation and stabilisation.

Walter Benjamin, who visited Moscow in the winter of 1926–1927, noted that the slowing down of the revolutionary tempo sat awkwardly alongside the drive to educate a new revolutionary generation:

> An attempt is being made to arrest the dynamic of revolutionary progress in the life of the state—one has entered, like it or not, a period of restoration while nonetheless wanting to store up the revolutionary energy of youth like electricity in a battery.[20]

Benjamin described a fresh political mood turning stale. He also detected a waning of enthusiasm in the revolutionary generation, which he described as a process of premature ageing:

> The generation that was active in the civil war years is growing old in vitality, if not in years. It is as if stabilisation had admitted to their lives the calm, sometimes even the apathy that is usually brought about only by old age.[21]

Benjamin suggested that the once youthful vigour of the revolution itself had been lost. A contradiction between transformation and consolidation also animated Luria's theorisations of the child, but unlike Benjamin he framed the maturation of both individuals and civilisation in positive terms. Luria aligned ageing with progress, rather than seeing it as a form of deterioration, detachment and waning strength. In the context of a discussion of nineteenth-century pedagogy, Benjamin declared that 'the adult was the ideal in whose image the educator aspired to mould the child'.[22] This aspiration also implicitly structured Luria's work, though he also encountered children who, for social and experiential reasons, would struggle to attain the ideal final destination.

[20] Benjamin, *Moscow Diary*, ed. by Gary Smith and trans. by Richard Sieburth (Cambridge, MA: Harvard University Press, 1986), p. 53.

[21] Benjamin, *Moscow Diary*, p. 45.

[22] Walter Benjamin, 'Toys and Play' (1928) in *Selected Writings*, vol. 2, ed. by Michael W. Jennings (Cambridge, MA: Belknap Press of Harvard University, 1999), pp. 117–121, p. 118. On the paradigm of development and the figure of the 'normal' child, see André Turmel, *A Historical Sociology of Childhood: Developmental Thinking, Categorization, and Graphic Visualization* (Cambridge: Cambridge University Press, 2008).

The Newborn Baby: From Organic Passivity to Historical Activity

'Imagine a man in whom all the links connecting him with the environment are cut off one after another; he turns out to be completely isolated from the world, a lone person amidst the world of things that do not exist for him.'[23] This is how Luria characterised the newborn baby in 'The Child and its Behaviour', the chapter he contributed to his and Vygotsky's book *Studies on the History of Behaviour: Ape, Primitive and Child* (1930). He also likened the infant to an anchorite monk. He began by attacking those who would present the child as an adult in miniature. This representational fallacy, he wrote, is typical of 'primitive' people and can also be observed in the history of art, which is littered with images of the baby Jesus whose bodily proportions resemble those of a grown man. Luria insisted instead that the child is a 'very special type of creature, who qualitatively differs from an adult and whose laws of life and activity should undoubtedly be studied with particular attention'.[24]

The baby's initial stage of 'irradiated arousal' is only gradually replaced by more stable perceptions.[25] At the beginning of life, babies perceive their external environments as if through 'a shroud', as 'chaotic, blurred' and as lacking 'any distinct forms'.[26] Luria claimed that young children continue to be solitary, passive and organic, concerned only with their own comfort. Young infants are, he says, 'parasitic'.[27] They have yet to encounter the obstacles presented by reality and seek only to satisfy their immediate

[23] Luria, 'The Child and Its Behaviour' in Vygotsky and Luria in *Studies on the History of Behaviour: Ape, Primitive and Child*, ed. and trans. by Victor I. Golod and Jane E. Know (Hillsdale, NJ: Lawrence Erlbaum, 1993), pp. 140–231, p. 145. According to the book's introduction 'The Child and Its Behavior' was written solely by Luria. See 'Translator's Introduction', pp. 1–35, p. 24. For the original, see A.R. Luriia, and Lev Vygotskii, *Etiudi po Istorii Povedeniia: Obez'iana, Primitiv, Pebionok* [*Studies in the History of Behaviour: Ape, Primitive and Child*] (Moscow: Gosudastvennoe Izdatel'stvo, 1930). On Luria's sole authorship of the chapter and Vygotsky's later remarks about the contradictions, it revealed in their distinction between biological and social development, see T.V. Akhutina, 'L.S. Vygotsky and A.R. Luria: Foundations of Neuropsychology', *Journal of Russian and East European Psychology*, 41, 3–4 (2003), 159–190, pp. 163–164.

[24] Luria, 'The Child and Its Behaviour', p. 145.

[25] Luria, 'The Child and Its Behaviour', p. 144.

[26] Luria, 'Experimental Psychology and Child Development' in *Selected Writings of A.R. Luria*, pp. 78–96. First published in *Nachnoe Slavoe*, 1 (1930), 77–97.

[27] Luria, 'The Child and Its Behaviour', p. 150.

wants. Similarly, speech for the young child is not used to communicate with others but is addressed to the self alone. Here, Luria cited research conducted by Jean Piaget into egocentric speech. Although he acknowledged universal characteristics in very young children, he noted suggestively that Piaget's conclusions are specific to the environment in which he conducted his experiments: 'Our children, developing in a different environment, may show different results.'[28] Elsewhere, Luria explicitly declared that children are full of 'shortcomings and inconsistencies',[29] outlining their inadequacies before asking, 'what, really, can we expect from such a creature?'[30] He was concerned with tracing the process by which humans overcome this initial stage of life through a series of dialectical transformations:

> in studies on children we are best able to observe directly how the thought of this creature, who has not yet become a part of complex social life, is being molded and what stages it passes through before it achieves the full-fledged status of the thought of a civilized, adult human being.[31]

As the baby grows into a child 'the world ceases to appear chaotic and unstructured'; discrete forms begin to emerge from the haze of undifferentiated colours and 'regular, organized, systematic forms of behaviour' begin to be established.[32] Luria explored this process through experiments involving a set of wooden blocks arranged in a square with one block placed outside it. He observed that very young children (1.5–2 years) did not perceive the form as a 'whole, dissociable figure'[33] and would pick up any block on the floor indiscriminately, whereas slightly older children would invariably pick up the block outside the square demonstrating that they perceived the blocks as a coherent figure.

[28] Luria, 'The Child and Its Behaviour', p. 155. Although later he notes that his colleagues repeated many of Piaget's experiments and came to the same conclusions, p. 167. In 'Paths of Development of Thought in the Child' (1929) he presents accounts of Piaget's experiments alongside his own.

[29] Luria, 'Paths of Development of Thought in the Child' in *Selected Writings of Alexander Luria*, ed. by Michael Cole (White Plains, NY: M. E. Sharpe, 1978), pp. 97–144, p. 99. First Published in Russian in *Estestvoznanie i Marksizm*, 2 (1929), 97–130.

[30] Luria, 'Paths…', p. 124.

[31] Luria, 'Paths…', p. 98

[32] Luria, 'Experimental Psychology', p. 83, p. 86.

[33] Luria, 'Experimental Psychology', p. 81.

Writing in 1931, Vygotsky outlined the methodological differences between Soviet pedology and its Western predecessors. Pedology, he argued, had not survived in the West, due to the inability of its practitioners to meaningfully combine approaches from various disciplines. The child, Vygotsky claimed, is a unique object and cannot be understood by falsely separating the physical from the psychological, the external from the internal and the biological from the social. Faced with the child's complex developmental transformations, these Western pedologists were unable 'to consciously reflect the dialectical nature of child development'.[34] Vygotsky claimed that a new methodology was required to grasp the unity (but not the identity) of mental and physiological processes, of thinking and being, of appearance and experience and of the general and the particular. Western pedologists were able to identify links between different phenomena (e.g. between intellect and memory, education and nourishment) but 'deprived them of any internal integrity and structure' and instead described nothing more than disconnected particles; 'a heap of broken glass ... a pile of broken atoms'.[35] By adopting a dialectical methodology informed by Marx, Soviet psychologists could, Vygotsky claimed, overcome these deficiencies. Such an approach would not only describe how different psychological phenomena related to one another but also situate the child in its particular social, cultural and historical environment. According to Vygotsky, the dialectical psychologist should consider nothing in isolation.

Dialectics for Vygotsky was not only a scientific methodology for analysing children but also provided a model for understanding how the child itself developed over time. Vygotsky likened the transition from childhood to adolescence to the caterpillar's transformation into a butterfly. This qualitative transition combined creation with destruction, life with death: 'The complex merging of processes of dying and being born interwoven one with the other.'[36] Like the Communist state's understanding of historical and economic development, Vygotsky carved childhood into discrete stages with a dialectical transformation occurring at each phase of

[34] Lev Vygotsky, 'K voprosu o pedologii i smezhnykh s neiu naukakh' [On the Question of Pedology and Adjacent Disciplines], *Pedologiia*, 4, 3 (1931), 53–58, p. 58.

[35] Vygotsky, 'K Voprosu...', p. 58.

[36] Lev Vygotsky, 'Pedology of the Adolescent', *The Collected Works of Lev Vygotsky, Vol. 5, Child Psychology*, trans. by Marie J. Hall (New York, NY: Plenum Press, 1998), pp. 3–186, p. 14.

transition.[37] Vygotsky described human life as a parallel to historical development as an 'ascent up the stages of development' in which stable periods were punctuated by crises.[38] He explicitly likened this to revolutionary historical processes. During stable periods 'development proceeds as if underground',[39] but then a crisis erupts at which point 'development takes on a stormy, impetuous and sometimes catastrophic character that resembles a revolutionary course of events'.[40] As Carolyn Steedman discusses in the context of Western European psychology, 'the child's developing body and mind could be understood as the epitome of a more general historical progress', though in Vygotsky's case the model of historical progress being mapped onto childhood development was distinctly Marxist-Leninist as understandings of historical progress are themselves historically specific.[41] In his delineation of the different stages of childhood, Vygotsky did not explicitly align the stages of human development with particular historical epochs (e.g. by comparing ancient Greece to infancy as Marx had done).[42] Rather than mapping the content of thought across history onto individual human life, the analogy was instead intended to describe the *form* of developmental transition, characterised by 'abrupt and major shifts and displacements, changes and discontinuities'.[43] The process he described was not smooth but uneven: 'the development of the child is a dialectical

[37] See Vygotsky, 'The Problem of Age' (1932–1934) *The Collected Works of L.S. Vygotsky, Vol. 5*, trans. by Marie J. Hall, ed. by Robert W Riener (New York, NY: Plenum Press, 1998), pp. 187–205, p. 196. The Communist Party was also structured into successive age groups: Young Octoberists (8–10), Pioneers (10–14), Komsomol (14–28). After passing through these stages a person could finally become a Communist proper. See Stanislav A. Pedan, *Partiia i Komsomol, 1918–1945* [*Party and Komsomol*] (Leningrad, 1979).

[38] Vygotsky, 'The Problem of Age', p. 188.

[39] Vygotsky, 'The Problem of Age', p. 188.

[40] Vygotsky, 'The Problem of Age', p. 191.

[41] Carolyn Steedman, *Strange Dislocations: Childhood and the Idea of Human Interiority, 1780–1930* (Cambridge, MA: Harvard University Press, 1995), p. 85.

[42] Marx described ancient Greece as the childhood of humanity: 'Does not in every epoch the child represent the character of the period in its natural veracity? Why should not the historical childhood of humanity, where it attained its most beautiful form, exert an eternal charm because it is a stage that will never recur?' Karl Marx, *A Contribution to the Critique of Political Economy* (Moscow: Progress Publishers, 1970), p. 217. Luria does explicitly link children's thought processes to earlier stages of civilisation: 'Parallels taken from the history of thought only confirm that the ability to abstract from a concrete initial point … is a product of a later period. An enormous amount of time was necessary before thinkers could abstract from the concrete empirical postulates of Euclid.' 'Paths…', p. 113.

[43] Vygotsky, 'The Problem of Age', p. 191.

process in which a transition from one stage to another is accomplished not along an evolutionary, but along a revolutionary path.'[44]

This development was not, however, seen to occur outside of history. As outlined in his critique of Western pedology, organic and historical factors were said to intersect. His emphasis on history differentiates this model of developmental stages from that proposed by Piaget.[45] Although certain developments were considered innate and biological (e.g. teething or puberty) Luria and Vygotsky also insisted that these could not be understood in isolation from social factors. Luria demonstrated this by way of an agricultural metaphor:

> At first, the earth produced as much as the natural conditions (its properties, weather, germination of random seeds) permitted; with cultural management and new conditions—fertiliser, development of better tools and maintenance-it began to yield a far greater harvest, and gradually (under prolonged cultural [*sic*]) the earth itself changed and proved to have adapted to maximum production.[46]

Just as Soviet agricultural policies aimed to apply technological innovations to the natural material of the earth in order to yield vast grain outputs, Soviet psychologists hoped that the natural material of the child could be enhanced by cultural means.

Luria and Vygotsky celebrated artificial accretions or 'auxiliary means' that extend human capacities. Tools and language were central to their account of human development. Indeed, Luria described a child's

[44] Vygotsky, 'The Problem of Age', p. 193. On the unevenness of development, see also Luria, 'Experimental Psychology…', p. 95. Luria similarly notes that the child's 'cultural development is often uneven, and experiments indicate that traces of primitive thinking often show up even in quite developed children', 'Experimental Psychology…', p. 95.

[45] Piaget's key works from this period are: Jean Piaget, *The Language and Thought of the Child* (London: Routledge and Kegan Paul, 1926), *The Child's Conception of the World* (London: Routledge and Kegan Paul, 1928) and *The Moral Judgment of the Child* (London: Kegan Paul, Trench, Trübner and Co., 1932). A discussion of the ways in which Vygotsky's theories were posthumously mobilised against Piaget's model by American and British psychologists in the post-war years is beyond the scope of this book. Piaget and Luria corresponded and were on friendly terms and in his works from the 1920s Luria tended to cite Piaget uncritically. In some comments on Vygotsky, Piaget refers to Luria as a friend and outlines some of the divergences between his theories and those of his Soviet contemporaries. See Jean Piaget, *Comments on Vygotsky's Critical Remarks* (Cambridge, MA: MIT Press, 1962).

[46] Luria, 'The Child and Its Behaviour', p. 212.

'entrance into life' not as its birth, but as its initiation into the cultural world.[47] The passive infant transforms into an active subject. Luria, drawing on Engels, emphasised that the formation of 'advanced' dialectical thinking came about through the interactions between people and their environment. Individual development could therefore not be understood in strictly natural terms:

> No development—that of the child included—in the conditions of modern civilized society can be reduced merely to the development of natural inborn processes and the morphological changes conditioned by the same; it includes, moreover, that social change of civilized forms and methods which help the child in adapting itself to the conditions of the surrounding civilized community.[48]

He was concerned with the development of capacities that allowed the subject to adapt to but also exert an influence on their external surroundings. For Luria, human subjects and external objects were mutually constituted: only through this transformative encounter could human thought ascend to a 'higher' stage. The dialectic between people and their surroundings constituted the basis of historical progress and precipitated the development of dialectical thinking:

> What traits have we become accustomed to seeing in behaviour guided by dialectical thinking? A keen regard for the real, the ability to take into account all sorts of changing conditions. So as to be able not only to adapt to the real world but also to predict its dynamics and to adapt it to oneself; a considerable plasticity and flexibility of behaviour that enable one to make use of different devices and different means, depending on the situation; and, finally, the definitive rejection of all pat, ossified forms of behaviour—these are the traits of behaviour that best reflect the dialectical method of thinking.[49]

Luria concluded this description of 'advanced' thought by remarking that 'it is just these traits that we do not observe in the behaviour of the small

[47] Luria, 'The Child and Its Behaviour', p. 145.
[48] A.R. Luria, 'The Problem of the Cultural Behaviour of the Child', *Journal of Genetic Psychology*, 35 (1928), 493–506.
[49] Luria, 'Paths…', p. 99.

child'.[50] His work with children was therefore directed at encouraging the development of this mode of perception from the youngest possible age.

According to Luria, the baby adapts to its environment through experience. The senses develop in tandem with memory. Eventually the 'accidental, separate and fluctuating scraps' of jumbled perceptions are replaced by more stable and complete pictures,[51] but as the external world becomes gradually more orderly, it begins to merge with the child's inner world: 'In perceiving the world holistically, the small child at the same time often loses the boundary separating reality from fantasy, the present from the past, what exists from what is desired.'[52] As a result, the child continues 'to hatch up unusually vivid fantasies' in order to live between two worlds.[53] For Luria, the transition to adulthood involved developing the ability to keep these two realms firmly apart. It is only then that the individual will be able to exert 'an efficient influence on the world'.[54] A process of cultural reconstruction must occur for the child to shed fantasies and ascend to the adult stage, adapted to (and therefore capable of adapting) the external world.

Paradoxes of the Vanguard

In *The Ignorant Schoolmaster* (1991), Jacques Rancière addresses the contradictory relationship of adult to child. Drawing on the example of a pedagogical experiment undertaken in the aftermath of the French Revolution, he asserts the possibility of learning between equals. Rancière describes learning as a tussle between stultification and emancipation, the stultified and the emancipated. For Rancière, assuming the inferiority of any other person will result in stultification, instead emancipatory learning must proceed from the assumption that 'everyone is of equal intelligence'.[55] Equality is understood as a practice in the present rather than a concept to be realised in the future. Rancière dispenses with a hierarchical relation that assumes a pairing of ignorance to knowledge. All people are assumed to have an equal capacity to learn, which, for Rancière, trumps the

[50] Luria, 'Paths…', p. 99.

[51] Luria, 'The Child and Its Behaviour', p. 148.

[52] Luria, 'The Child and Its Behaviour', p. 149.

[53] Luria, 'The Child and Its Behaviour', p. 149.

[54] Luria, 'The Child and Its Behaviour', p. 151.

[55] Jacques Rancière, *The Ignorant Schoolmaster: Five Lessons in Intellectual Emancipation*, trans. by Kristin Ross (Stanford, CA: Stanford University Press, 1991), p. 101.

disparities between what different people happen to know. The relation between stultification and emancipation he describes might be aligned with the confrontation of reproduction with revolution and routine with rupture traced in this chapter. Rancière is primarily concerned with how to address the asymmetrical power relationship between student and teacher. The pedagogical parity he describes was crucially, he insists, antithetical to Enlightened notions of progress: 'Equality was not an end to attain, but a point of departure, a supposition to maintain in every circumstance.'[56] For Rancière, liberation cannot be founded on oppression.

A tension between stultification and emancipation is played out in Luria's and Vygotsky's writings on childhood. They did not share the ideas expressed by some of their more radical Soviet contemporaries that adults should follow the lead of their more authentically Communist progeny. For Aron Zalkind, for example, children would ideally form miniature cadres of dedicated revolutionaries untarnished by the former 'stagnant way of life', setting an example to the older generation rather than meekly following its lead: 'children—enthusiasts, activists, gripped by collective feeling, carry into musty family life new principles of revolutionary ideology … a future is being formed in its embryonic elements through the children's efforts.'[57] Adults were more central to Luria and Vygotsky's accounts of childhood, playing a key role in introducing children to the social world, a world of words and objects organised in particular ways, as Luria discussed in a 1936 article:

> in cultural development the relations between man and reality are the motive power of progress. For human beings these relations are of a social nature. The higher psychological functions are developed by social contact, and in the first place by the intercourse of the child with adults.[58]

[56] Rancière, p. 138.

[57] Aron Zalkind, 'The Pioneer Youth Movement as a form of Cultural Work among the Proletariat' in *Bolshevik Visions: The First Phase of the Cultural Revolution in Soviet Russia*, ed. by William G. Rosenberg (Ann Arbor: Ardis, 1984), pp. 347–354, p. 351. First published in Russian in *Vestnik truda*, 3, 40 (1924), 107–116.

[58] Luria, 'The Development of Mental Functions in Twins', *Character and Personality*, 5, 1 (1936), 35–47, p. 36. See also, Luria, 'The Variability of Mental Functions as the Child Develops (Based on a Comparative Study of Twins), *Soviet Developmental Psychology*, p. 65. First published as: 'Ob izmenchivosti psikhicheskikh funkstii v protsesse razvitiia rebenka', *Voprosy psikhologii*, 3 (1962), 15–22. This piece discusses research N.G. Morozova conducted with twins in the 1930s. In a letter to Michael Cole Luria indicated her findings remained unpublished: 'As to your question concerning my early work on twins I asked N.G.

In their experiments, they traced how children gradually learnt to appeal to adults for assistance: 'the road from object to child and from child to object lies through another person.'[59]

Vygotsky's 'Play and its Role in the Mental Development of the Child' suggested that in play the child operates according to rules of his or her own devising, rather than relying on the external rules imposed by parents or guardians. Children become conscious directors of their own activities. In 'Tool and Symbol in Child Development' (c. 1930), Luria and Vygotsky similarly analyse how the child learns to master its environment through the use of speech, a capacity which, they argue (recalling their critiques of Gestalt psychology discussed in Chap. 2), qualitatively separates children from apes. With the aid of speech, eventually internalised as thought, the child can plan and organise its activities. Narrating action is not a passive activity but plays an active role in the performance of practical tasks. Symbolic activity and practical activity form a 'complex unity' in the human.[60] For Luria and Vygotsky, the emergence of this 'newly born unit of perception' marks the beginning of human life proper; the beginning of a mode of cognition fundamentally distinct from the animal:

> At the moment when, thanks to the planning assistance of speech, a view of the future is included as an active agent, the child's whole operational psychological field changes radically and its behaviour is fundamentally reconstructed.[61]

Morozova (who is now 70!!) and she assured me that nothing of her work of these years was published. She will try to find the original ms., and if she will succeed, I shall send it to you.' A.R. Luria to Michal Cole, October 16, 1976, Michael Cole Personal Archives.

[59] Luria and Vygotsky, 'Tool and Symbol in Child Development' in *The Vygotsky Reader*, René van der Veer and Jaan Valsiner eds. (Oxford: Basil Blackwell, 1994), pp. 99–174, p. 116. Although Vygotsky lists this essay in the bibliography of *Myslenie i rech* (*Thinking and Speech*) in 1934, this text was not published in Russian until 1984. On the text's history see *Vygotsky Reader*, p. 170. An exhaustive comparison of the English and Russian versions is provided in D. Kellogg and A. Yasnitsky, 'The Differences between the Russian and English Texts of Tool and Symbol in Child Development—Supplementary and Analytic Materials', *PsyAnima, Dubna Psychological Journal*, 4, 4 (2011), 98–158.

[60] Luria and Vygotsky, 'Tool…', p. 113. Critiques of Vygotsky published in the 1930s emphasised that he did not think adult guidance was a pedagogical prerequisite, claimed he was too quick to dispenses with rules, reprimands and examinations, and drew attention to passages that argued for an 'anti-Leninist' abolition of the school as an institution.

[61] Luria and Vygotsky, 'Tool…', p. 122.

Through speech, the child masters both itself and its environment. Planning and intentionality are necessary to comprehend reality as more than 'splintered and peculiar'[62] sensations or as a 'hodgepodge of disrupted and disorganised' perceptions.[63] Language is also necessary for the child to meaningfully manipulate objects. Vygotsky argued that in play

> a child is always above his average age, above his daily behaviour; in play it is as though he were a head taller than himself ... in play it is as though the child were trying to jump above the level of his normal behaviour.[64]

For Vygotsky, the child takes on the position of its own educator. In the prescribed imaginary space of the game, the child learns to apply the order inherent to the existing adult world. Rules are central to Vygotsky's account of play, paving the road to working life: play 'permeates the attitude toward reality. It has its own inner continuation in school instruction and work (compulsory activity based on rules)'.[65] Through play the child learns to derive pleasure from the performance of unpleasurable activities, internalising the 'higher' structures of thought of the adult. Although adults do not directly guide the child here, their internalised presence is indispensable. The function of play remains assimilative and is premised on a progressive understanding of development understood as a process of ascent towards adulthood.

The period during which the child learns to play is preceded by a period of total dependence on the adult: 'Definitely everything in the behaviour of the infant is intertwined and interwoven into the sociable.'[66] In Luria's and Vygotsky's accounts, adults, who children initially mistake for objects, appear as the custodians of reality. A child's relation to reality is 'always refracted through a prism of relations with another person'.[67] Adults are responsible for introducing children to language and social existence. In the intellectual biography *Understanding Vygotsky*, René Van der Veer and Jaan Valsiner summarise the relation of child to adult in Vygotsky's work in the following terms:

[62] Luria and Vygotsky, 'Tool...', p. 125.

[63] Vygotsky and Luria, 'Tool...', p. 112. Here they are comparing the child to the aphasic person who, they argue, is similarly incapable of coherently uniting their perceptions.

[64] Vygotsky, 'Play', pp. 95–96.

[65] Vygotsky, 'Play', p. 98.

[66] Vygotsky, 'Infancy', *The Collected Works of L.S. Vygotsky, Vol. 5*, pp. 207–241, p. 215.

[67] Vygotsky, 'Infancy', p. 216.

> Children should not be left to themselves when acquiring new knowledge and wisdom, as this was equivalent to not educating them at all, leaving their development to the obnoxious forces of the 'street'. Bringing up children necessarily means restricting their freedom of action, Vygotsky reasoned, sometimes in the interest of the child himself, sometimes in the interest of the collective. The role of parents and teachers is and should be enormous: in forming part of the children's environment and organising this environment they will steer the child's mental development to a considerable extent.[68]

Similarly, in Luria's experiments with twins, adult educators play a key role in directing activities. The children who tended to emerge as the most 'advanced' were the ones who received most instruction from educated adults.

Soviet Toys

Vygotsky and Luria's *Ape, Primitive and Child*, first published in 1930 of which Luria's 'The Child and Its Behaviour' is a chapter and of which Vygotsky was later critical,[69] explores three distinct forms of psychological development: evolutionary, cultural and individual. In their introduction, the authors explained that in each case they were interested in describing a qualitative, dialectical transformation that concerns the development of 'auxiliary means'. The term comes from a passage by Francis Bacon, which forms the book's epigraph: 'A bare hand and a mind by itself are worth not that much: everything is performed with the help of tools and auxiliary means.'[70] In the context of childhood, they defined toys and games as tools of this kind. Play plays a key role in Luria's and Vygotsky's theorisations of the child.

Luria discussed play in the entry he wrote on 'Imagination' for the 1929 edition of the *Great Soviet Encyclopedia*, edited by his erstwhile

[68] René van der Veer and Jaan Valsiner, *Understanding Vygotsky: A Quest for Synthesis* (Oxford: Blackwell, 1991), p. 53

[69] Akhutina, 'L.S. Vygotsky and A.R. Luria: Foundations of Neuropsychology', pp. 163–164.

[70] Vygotsky and Luria, *Ape, Primitive and Child*, p. 40. In the English translation this is included as an epigraph to the opening chapter, whereas in the Russian it is given a separate page at the beginning of the whole book. See L.S. Vygotsky and A.R. Luria, *Etiudi po Istorii Povedeniia: Obez'iana, Primitiv, Pebionok* (Moscow: Gosudastvennoe Izdatel'stvo, 1930), p. 7.

Russian Psychoanalytic Society colleague Otto Schmidt, which began by identifying two distinct forms of imagination:

> passive or reproductive imagination, which consists of recalling images of past experience in the memory without substantially changing them, and active or productive imagination, a different process which involves combining the results of experience [to create] new images.[71]

Imagination is said to play a role in all intellectual, practical and creative activities, as well as in dreaming and in waking fantasy. Imagination relies on the capacity to 'spontaneously call forth images that are not drawn directly from environmental stimuli'.[72] Luria characterised imagining as a visual process that involves creating new images in the mind. Although this relies on memory it demands more than a 'mere repetition' of previous experience. The non-immediacy of imagination is not, however, irregular or anarchic but is governed by laws which, according to Luria, psychologists are capable of analysing and describing. He went on to discuss the centrality of the imaginative capacity in children, claiming that 'distinguishing reality from the imaginary comes only at a certain stage of development'.[73] For Luria, this capacity to differentiate develops through play. In his account play is simultaneously reproductive and creative; it participates in children's ascension to a higher cognitive level. The function of play is not to invent alternate realities, new worlds or original narratives, however: games are a rehearsal for adult life. Luria differentiated play from work in its relationship to the realm of necessity. Unlike an adult worker the child at play does not need to plan and 'what is real and what is illusory are blended'.[74] Luria may have criticised representations of children as tiny adults and was explicit in differentiating play from work, yet he was dedicated to ensuring these peculiar proto-human creatures adopted the features of their adult progenitors as swiftly as possible.

In November 1931 the first exhibition of children's toys was held in Moscow. In stark contrast to the simple and traditional objects hewn from

[71] Luria, 'Voobrazhenie' [Imagination], *Bol'shaia Sovetskaia Entsiklopediia, tom 13*, ed. by Otto Schmidt (Moscow: Aktsioneroe Obshchestvo Sovetskaia Entsiklopediia, 1929), pp. 120–121.

[72] Luria, 'Voobrazhenie' [Imagination], p. 121.

[73] Luria, 'Voobrazhenie' [Imagination], p. 121.

[74] Luria, 'Paths…', p. 127.

'wood, clay, bone, textiles, paper, papier-mâché'[75] that Walter Benjamin described after visiting the toy museum in Moscow, this presented a vision of industrialised Soviet society in miniature, with traces of the past erased. The exhibition included a miniature model of the Dnipropetrovsk dam (one of the showpieces of the First Five Year Plan) and a tiny conveyor belt at which two shock brigades of Pioneers were set to work assembling toys. Like a living maquette of their Stakhanovite adult counterparts, the children were set in competition with one another; the importance of maximum efficiency and a love of work were to be instilled at the earliest stages of life.[76]

In 1935, the Pedological and Pedagogical Laboratory published a study investigating children's play, using new Soviet toys; toys designed to prefigure a new way of life. It included dolls corresponding to the nationalities of the Soviet Union and different class groups and games based on the *kolkhoz*, the Park of Culture and Rest, public dining rooms, public laundries, factories and modes of transport (particularly locomotives, an archetypal symbol of progressive Soviet modernity). Construction, mastery of complex narrative and cooperation were the central goals of these toys.[77]

This emphasis on fostering efficiency and practicality was also evident in Luria's research. In 1934 Luria returned full time to Moscow from Kharkiv, where he conducted experiments with sets of twins of different ages at the Medico-Genetic Institute in Moscow, drawing on a card index in his possession listing all the twins in the Soviet Union.[78] Working with

[75] Benjamin, 'Russian Toys', p. 123. Benjamin also discusses the industrialisation of toy manufacture in 'The Cultural History of Toys' (1928), *Selected Writings, Vol. 2*, pp. 113–116.

[76] See Ella Winter, *Red Virtue: Human Relationships in the New Russia* (New York, NY: Harcourt Brace and Co, 1933), p. 244. A history of Russian peasant toys published in 1933 ends with a chapter on contemporary toys made in the 'proletarian style', which include a crane and a pair of Red Army soldiers: N. Tseretelli, *Russkaya Krest'ianskaia Igrushka* [*Russian Peasant Toys*] (Moscow: Akademia, 1933). For an illustrated overview of the collection of the Zagorsk Museum of Toys, USSR Academy of Pedagogical Sciences, see G.L. Dain, *Russkaia Igrushka* [*Russian Toys*] (Moscow: Sovietskaia Rossiia, 1987).

[77] E. Molozhavaia, *Siuzhetnaia Igrushka—Tematika i Oformlenie* [*Toys—Theme and Design*] (Moscow and Leningrad: Vsesoyuznoe Kooperativnoe Ob'edlenennoe Izdatel'stvo, 1935). An earlier collection of psychological investigations into children's relationships with toys overwhelmingly relied on traditional toys: N.A. Rybinkov ed., *Rebionok i igrushka* [*Children and Toys*] (Moscow: 192–).

[78] Homskaya, p. 28.

both fraternal and identical twins at different ages, these investigations aimed to probe the relationship between natural and social factors in childhood development to establish 'the ratio of hereditary to environmental influence'.[79] 'The Experimental Development of the Constructive Activity—on the Differential Training of Identical Twins' (1936), co-authored with A.N. Mirenova, is based on the analysis of three pairs of seven-year-old identical twins. Each pair of twins was separated to create two groups. The first group (E) were trained to passively copy figures composed of wooden blocks. The second group (M) were also trained to reproduce the composition but were only shown the outline of the final shape, rather than the positioning of each individual block. The twins were then given a series of tests that required them to reproduce different figures on paper or to identify geometrical shapes in a puzzle. The second group was shown to be more capable of accurate reproduction than the first, even 18 months after the initial training has occurred. The capacity to accurately reproduce drawings and constructions was deemed the most desirable outcome of the experiment. Luria and Mirenova claimed that creativity was required to accomplish this task; it was not a passive or automatic activity.[80] In tune with Luria's understanding of imagination, this is a definition of creativity which allows for the reproduction of existing forms but does not encourage children to invent their own constructions or images from scratch.

In another paper, based on the same experiments, Luria was explicit that '*ad hoc* creative play' is not conducive to learning and warned against allowing children's activities to 'degenerate into simple spur-of-the-moment play'.[81] Instead, he advocated games that are goal oriented but that do not simply involve passive reproduction. Although he conceded that that a form of play based around 'free building' is more entertaining for the child, he argued that it has no pedagogic function as the child

[79] Luria, 'The Development of Mental Functions in Twins', *Character and Personality*, 5, 1 (1936), 35–47, p. 35.

[80] A.R. Luria and A.N. Mirenova, *Eksperimentalnoie rasvitiye konstrukivnoi deiatelnosti* [*The Experimental Development of Constructive Activity*] (Moscow: Mediko-Geneticheskogo Institut, 1936), pp. 487–505.

[81] Luria, 'The Development of Constructive Activity in the Preschool Child' (first published in 1948 but based on the experiments with twins undertaken with Mirenova in the 1930s) in *The Selected Writings of Alexander Luria*, pp. 201–202. First published in *Voprosy psikhologii renenka doshkol'novo vozrata* [*Problems in the Psychology of the Preschool Child*] (Moscow and Leningrad: APN RSFSR, 1948), pp. 34–64.

proceeds 'without an analysis of the objective properties with which [he or she] is playing'.[82] In this scenario chaos reigns: a wooden block might become a cow, a dog or a tree. The child manipulates everything according to its whim and does not learn to accommodate 'realistic' material constraints. The 'creative construction' involved in reproducing a structure from a model developed by Mirenova is, by contrast, deemed goal oriented and constrained: 'the child is always obliged to remain within the framework of constructive activity and runs no risk of drifting into a merely nominal labeling of the elements'.[83] Luria wanted to replace the fragmented perceptions of the child with the unified coherence of the adult. Giving free rein to the imagination might be fun, but Luria insisted sternly, it has no practical social utility and would not encourage the child to progress intellectually. He again discussed the two groups of twins split into (M) and (E) noting that the former learnt how to plan more successfully than the latter who remained 'almost totally unsystematic', incapable of identifying their mistakes.[84] The former group were then found capable of mastering other tasks requiring 'complex visual analysis',[85] for example, mirroring the gestures of the experimenter. These experiments demonstrated that two twins previously almost identical in their capacities could, through different forms of education, diverge substantially; culture could supersede biology.

In Luria's twin experiments children are praised for creating things that corresponded to reality: reproducing an existing wooden structure is better than inventing a structure from scratch, making a plasticine table with four legs is better than making one with two. One child was praised for busying himself with the construction of a complex metro system including tunnels, stations and trains. He later struggled to sleep as he was 'thinking all the time about the metro'.[86] His fierce resistance to the

[82] Luria, 'Development of Constructive Activity…', p. 199.

[83] Luria, 'Development of Constructive Activity…', p. 201.

[84] Luria, 'Development of Constructive Activity…', pp. 207–208.

[85] Luria, 'Development of Constructive Activity…', p. 222.

[86] Luria and F.Ia. Yudovich, *Speech and the Development of Mental Processes in the Child*, ed. and trans. by Joan Simon (London: Penguin, 1971), p. 90. The first English edition was published in 1959 based on the Russian publication from 1956: *Rech' i razvitie psikhicheskikh protsessov u rebenka* (Moscow: APN RSFSR, 1956). The book's translator Joan Simon noted in a letter to the scientist Oliver Zangwill in 1975 that interest in the book had grown over time: 'In fact it has an interesting history—how it was ignored at first but demand has gradually spread'. Joan Simon to Oliver Zangwill, May 8, 1975, Brian Simon Archive, Institute of Education (IOE), DC/SIM/2/5.

structure being disturbed by a playmate was given as an example of his cognitive progress: he 'restored the building and returned to playing with it'.[87] For Luria, mastering such activities paved the way for adult life. Perhaps the child would one day work building a real Soviet Metro system.[88] In Luria's account of childhood development, language also played a key role in the construction of this pathway from disordered individualism to organised goal-oriented collectivity.

Language and Organisation

Luria and Vygotsky were both fond of citing the passage from *Capital vol. 1* in which Karl Marx distinguishes the spider from the weaver, the bee from the architect.[89] Although Marx claimed insects might 'put to shame' humans with their intricate constructions, the process of building is different for the human who 'raises his structure in imagination before he erects it in reality'.[90] For Marx, it is the ability to plan and imagine that distinguishes human labour from the activity of animals, and for Luria and Vygotsky it also distinguished 'lower' from 'higher' forms of thought. Abstract thinking for them was related to the capacity to plan, reproduce and organise. Abstraction in this definition was not connected to abstract forms of representation. In his analysis of children's drawings, for example, Luria deemed non-figurative scribbles a sign of underdevelopment. As with his experiments in Central Asia, it was the ability to think in concepts and to classify objects according to socially established criteria that represented the successful transition to a 'higher' stage of development. He did not celebrate the surreal or absurd.[91]

Speech and the Development of Mental Processes in the Child, co-written with F. Ia. Yudovich, was also based on Luria's research with twins in the early 1930s. The book focused on one pair of identical twins, Yura and

[87] Luria and Yudovich, *Speech*..., p. 91.

[88] The Moscow Metro was one of the major emblems of Stalinist modernity. It opened in 1935. Illustrated children's books celebrating the new construction indicate the extent to which knowledge of such projects were addressed to children. See Elizabeta Tarakhovskaia, *Metropoliten* (Moscow: Molodaia Gvardiia, 1933) and *Gotov! Rasskaz i stikhi o Metro* [*Ready! Stories and Poems about the Metro*] (Moscow: Detizdat, 1935).

[89] For a typical example, see Luria, *Speech*..., p. 87.

[90] Karl Marx, *Capital*, vol. 1, http://www.marxists.org/archive/marx/works/1867-c1/ch07.htm

[91] Luria, *Speech*..., pp. 96–97.

Liosha. Luria and Yudovich note that although the twins were 'good, cheerful, energetic, mischievous, friendly and affectionate', they are insufficiently advanced.[92] The twins had grown up playing together and developed a private language that was more simplistic or 'concrete' than that of most children their age. They had a very limited vocabulary, tended to distort words and the language they used tended to pertain to their immediate surroundings or situation. In the experiment the twins were first observed playing together, before being placed in a communal children's home where they were separated, with one receiving additional language lessons. The book traces the changes that occur in the twins' speech and behaviour as a result.

In one experiment, the twin who had received additional instruction is shown as being capable of grouping objects together according to various criteria (function, colour, etc.) and narrates his rationale for doing so as he proceeds, while the other remains slow and confused. In another experiment, the separated twins were both presented with images of animals with human apparel or attributes. Luria and Yudovich recorded their responses (Child A has received extra tuition unlike Child B):

> Yura (A):
>
> 'Does this happen?' 'It does' 'What is the cat doing?' 'The cat is playing' 'Can a cat really play on the violin?' 'No' 'Then does this happen?' 'No'
>
> Liosha (B):
>
> 'Is this drawing right?' 'It is' 'Cat a cat really dance like this?' 'No' 'Then is the drawing right or not?' 'Right' 'Have you seen a cat play on a balalaika?' 'No' 'Then is this drawing right or not?' 'Right' 'But can a cat play on a balalaika?' etc.[93]

This example is used to demonstrate the development of a 'theoretical' attitude. The first child is deemed more advanced than second because he was able to identify that a cat cannot actually play a musical instrument. As with his experiments with wooden blocks, Luria is clear that imagination is valued insofar as it relates to the capacity to plan ahead, and to construct and enact realistic goals for the future. A clear division between fantasy and reality needs to be instilled for the healthy development of the child.

[92] Luria, *Speech*..., p. 40.
[93] Luria, *Speech*..., p. 100.

Children should not persist in thinking that cats can really play the balalaika.

In the 1920s and early 1930s, there were conflicting ideas in the Soviet Union about the fantasy life of children. Particularly fierce debates focused on the issue of fairy tales (traditional Russian folktales known as '*skazki*'). The mainstream Party line initially advocated an attack on children's literature that departed from everyday life with one tract proclaiming: 'There is nothing to justify cats dressed up in hats or dogs baking pies; these fairground images add nothing positive to a child's emotions, nor do they inculcate clear ideas about animal behaviour.'[94] Dorothy Thompson, who visited the Soviet Union in 1928, recalled being informed by a kindergarten teacher of a proletarian re-writing of Cinderella which saw the monarchy overthrown with the justification that 'we do not want the energies of this coming generation dissipated by fantastic dreams about things which have no relation to our life and its problems'.[95] Constructivist artists produced works aimed at children and children's literature remained a haven for avant-garde writers and artists into the 1930s. But though these works were illustrated with dynamic, geometric illustrations full of bright colours and off-kilter figures, the narratives tended to be grounded in everyday life. The communal kitchens, automatic cooking appliances, zooming locomotives and slick factories that appeared in Soviet children's literature of the NEP period may have been as fantastical as any fairy tale, but they were intended as prototypes for the future.[96] As such, they might be aligned with the form of imaginative capacity discussed by Luria.

Luria approvingly cited the influential children's author Nikolai Chukovsky in 'The Child and His Behaviour'. Chukovsky's work came under attack in the late 1920s, but he eventually emerged triumphant, surviving the purges to become one of the Soviet Union's most prominent

[94] Cited in Catriona Kelly, *Children's World: Growing Up in Russia, 1890–1991* (New Haven, CT and London: Yale University Press, 2007), p. 74.

[95] Dorothy Thompson, *The New Russia* (New York, NY: H. Holt and Co., 1928), p. 233.

[96] For a discussion of the Soviet avant-garde's work for children, see Evgeny Steiner, *Stories for Little Comrades: Revolutionary Artists and the Making of early Soviet Children's Books* (Seattle, WA: University of Washington Press, 1999) and Alla Rosenfeld, 'Does the Proletarian Child Need a Fairytale?' *Cabinet*, 9 (2002–2003). For examples of Soviet children's book illustrations, see *Inside the Rainbow: Russian Children's Literature 1920–1935*, ed. by Julian Rothenstein et al. (London: Redstone Press, 2013).

children's authors.[97] Although socialist realism was established as the dominant Soviet literary genre at the Soviet Writers' Congress in 1934, this, somewhat counter-intuitively, coincided with the official reintroduction of traditional fairy tales for children.[98] Chukovsky was to play a prominent role in the eventual denunciation of Pedology later in the decade. His *Little Children*, first published in 1928 and reissued numerous times (under the title *From Two to Five*), presented a collection of the utterances of children, characterising them as 'apt, beautiful and natural'.[99] He praised children's joyful and inquisitive minds, delighting in their malapropisms and incorrect suppositions.

Chukovsky attacked early Soviet educators for not recognising the value of teaching poetry to children, quoting from a father's letter that he said was typical of attitudes of the late 1920s:

> Shame on you Comrade Chukovsky, for filling the heads of our children with all kinds of nonsense, such as that trees grow shoes. I have read with indignation in one of your books such fantastic lines as "Frogs fly up in the sky/Fish sit in the fishermen's lap/Mice catch cats/And lock them up in/Mousetraps". Why do you distort realistic facts? Children need socially useful information and not fantastic stories about white bears who cry cock-a-doodle-doo.[100]

Chukovsky railed against the crushing of children's creative impulses claiming that an overzealous attachment to realism was restrictive and 'risks fading the colour out of the child's speech, making it anaemic and devitalised, killing in it its wonderful childishness and inflicting a permanent

[97] Chukovsky's story 'Crocodile' was denounced by Nadezhda Krupskaya in *Pravda* as a 'bourgeois muddle' on 1 February 1928. On 17 January 1928, Chukovsky wrote in his diary that 'anthropomorphism' was considered the greatest crime of Soviet children's literature and that he hoped to write a defence of the fairy tale. See, Kornei Chukovsky, *Diary, 1901–1969*, trans. by Michael Henry Heim (New Haven, CT: Yale University Press), p. 217.

[98] For an overview of the early debates in Soviet children's literature, see Marina Balina, 'Creativity Through Restraint: The Beginnings of Soviet Children's Literature' in *Russian Children's Literature and Culture*, ed. by Marina Balina and Larissa Rudova (New York, NY: Routledge, 2008), pp. 3–17.

[99] Kornei Chukovsky, *From Two to Five*, trans. and ed. by Miriam Morton (Berkeley: University of California Press, 1963), p. 4. This English translation is a heavily edited version of a later edition. The discussions of pedology are not included in the Russian first edition when Chukovsky himself was under attack. See *Malen'kie Deti* [*Little Children*] (Leningrad: Krashaia Gazeta, 1928).

[100] Chukovsky, *From Two to Five*, pp. 89–90.

harm'.[101] Children, he argued, have an innate aversion to the established order of things, as such they delight in stories and games that enact a 'violation of reality': 'the child is attracted to that topsy-turvy world where legless men run, water burns, horses gallop astride their riders, and cows nibble on peas on the top of birch trees.'[102] Unlike Luria, who maintained that young children confused reality and fantasy and were unable to distinguish between images and objects—'the child would try to grab an apple in a picture'[103]—Chukovsky insisted that children are well aware of the distinction between the two: a child does not eat the sand pie that it bakes on the beach.[104]

Chukovsky's diatribes against people intent on sapping the 'creative genius of the people'[105] were explicitly aimed at pedologists who he accused of making a 'banal fetish' of practicality.[106] He recounted that in 1929 he visited a children's home where a fairy-tale book was snatched from him by a man in uniform:

> the pedologists are worried and tremble at the thought that children will actually believe that shoes grow on trees. Some children are so suspicious of everything—even the most poetic, that is, the most unreal—that everything beyond the limits of the everyday and the ordinary they consider a bold-faced and senseless fabrication.[107]

For Chukovsky, however, an inversion of reality always existed as a counterpart to reality itself, presenting an image of the world turned upside down therefore served to enforce the child's orientation to the existing state of things. Ultimately, Chukovsky and the pedologists he attacked shared assumptions about childhood perception and learning processes. Similarly, both Chukovsky and Luria were ultimately concerned with

[101] Chukovsky, *From Two to Five*, pp. 17–18.

[102] Chukovsky, *From Two to Five*, p. 95, p. 96.

[103] Luria, 'Experimental Psychology…', p. 85.

[104] Chukovsky's assumption is further challenged by R.G. Vilenkina's pedological experiments. She asked adolescent workers and apprentices to deposit anonymous notes in a designated box. One asks—'Do people live on Mars?' suggesting that Soviet fantasies were indeed mistaken for reality by some contemporary citizens. See R.G. Vilenkina, 'K kharakteristike nasroenii rabochego podrostka' [On the Characteristic Mental Features of the Adolescent Worker], *Pedologiia*, 1 (1930), 81–97, p. 88.

[105] Chukovsky, p. 112.

[106] Chukovsky, p. 125.

[107] Chukovsky, pp. 124–125.

teaching the child to orient itself to the established structures of the world.[108]

Luria, like Piaget who famously observed his three children, made notes on his daughter Elena's physical development, affective responses and perceptions of colour and sound as she grew up.[109] His observations about her linguistic development emphasised the relation between speech and the gratification of need (e.g. he notes that she initially communicated in imperatives indicating basic wants). He also gave examples of infantile misunderstanding, relating an occasion when she mistakenly called a puddle a 'shame' as this was what her nurse always exclaimed when the child wet herself. Luria described this as a 'comical event' but, like Chukovsky, interpreted the incident as a mistake to be corrected.[110] Walter Benjamin also documented his son Stefan's language usage, collecting his childish aphorisms and idiosyncratic formulations in a special notebook. In her introduction to a translation of the surviving fragments of this collection, Esther Leslie differentiated Benjamin's observations from the more systematic descriptions of Piaget (who in this respect was much closer to Luria), noting that his project was

> not directed toward a documentation of the acquisition of language, but rather toward specific aspects of infantile thinking and speaking—the detours in which it goes astray. … For him, linguistic blunders and misunderstandings do not find their meaning in being corrected.[111]

For Benjamin, Stefan's associations and correspondences cracked open the adult world to reveal relations between things obscured by more habitual or normative perceptions.

In Luria's work, by contrast, language acquisition had an organising function: 'We regard speech as a powerful cultural tool that transforms

[108] A similar observation about the similarities between Chukovsky's approach and the approach of pedologists he aims to distinguish himself from is made in Karin Lesnik-Oberstein, *Children's Literature: Criticism and the Fictional Child* (Oxford: Oxford University Press, 1994), pp. 2–3.

[109] Sections from Luria's notebooks describing his daughter's development are included in Elena Luria, *Moi Otets*, pp. 92–99.

[110] Elena Luria, *Moi Otets*, p. 96.

[111] Esther Leslie, 'Opinions et Penseés: His Son's Words and Turns of Phrase' in *Walter Benjamin's Archive: Images, Texts, Signs* (London: Verso, 2007), pp. 109–112. For sections of the notebook, see pp. 116–150.

primitive, natural forms of behaviour into complex, cultural forms.'[112] The pre-linguistic child 'does not yet possess correct and adequate ideas about a whole order of things'.[113] Through naming the child begins to differentiate objects, classify entities in accordance with established systems and communicate in a more sophisticated manner: 'The child becomes aware that a particular object may play a different role in different situations; and his perception, refracted through this prism, acquires a more dynamic nature.'[114] For Luria this enabled the child's thinking to become 'more vital and practical'.[115] 'Dynamism' for Luria tended to be related to mastery of existing structures, rather than their subversion.

Luria clearly articulated that child psychology was valuable for its capacity to shed light on the transition from 'natural' baby to 'cultural' adult:

> in studies on children we are best able to observe directly how the thought of this creature, who has not yet become a part of complex social life, is being molded and what stages it passes through before it achieves the full-fledged status of the thought of a civilized, adult human being.[116]

His work with children was ultimately concerned with explaining the emergence of 'a cultured adult living in complicated industrial-cultural conditions'.[117] In the article in which he made this statement he mentioned that a child growing up in a city would think differently from a child growing up in a town but gave no more detailed sense of the specificities of Soviet cities and provincial areas at this moment in history. Though his experiments with twins were designed to identify the influence of 'nurture' over 'nature', they were laboratory bound and operated with a restricted definition of nurture pertaining primarily to differences in formal education. Yet the chaotic social conditions of Soviet life proved difficult for Luria and his collaborators to ignore and prior to conducting the experiments with twins he had also left the laboratory behind to conduct experiments with children who had grown up in

[112] Luria, 'Paths…', p. 133.
[113] Luria, 'Paths…', p. 124.
[114] Luria, 'Paths…', p. 137.
[115] Luria, 'Paths…', p. 142.
[116] Luria, 'Paths…', p. 98.
[117] Luria 'The Problem of the Cultural Behaviour of the Child', *Journal of Genetic Psychology*, 35 (1928), 493–506, p. 504.

different social environments. On Soviet streets Luria encountered young people who deviated from the ordained path to maturity due to their social experiences. How could orderly minds be forged in the midst of concrete devastation? And how could the continued existence of social and psychological disorder be made to fit with a progressive framework for understanding historical and individual development?

Children of the Revolution

In 'Old Forgotten Children's Books' (1924), Walter Benjamin outlined an understanding of childhood associated with the Enlightenment: 'if man was pious, good, and sociable by nature, it had to be possible to transform children, who were creatures of nature in its purest form, into the most pious, the best, and the most sociable beings of all.'[118] In her discussion of children's fiction, Jacqueline Rose similarly identifies a 'fetish of childhood' in Western European literature, which conceives of the early years of life as a period of lost innocence and unity.[119] In Enlightenment thinking, according to Rose, the child is likened to the 'primitive': 'They are connected by a fantasy of origins—the belief that each one represents an ultimate beginning where everything is perfect or can at least be made good.'[120] The child functions as a 'pure point of origin'.[121] In Jean-Jacques Rousseau's canonical *Émile* (1762), Rousseau declared human existence a degradation, disfiguration and deformation of the integral and harmonious nature created by God. Though he acknowledged returning to this whole and unmutilated beginning was impossible, Rousseau outlined an educational programme that attempted to adhere as closely as possible to the natural course: 'Observe nature and follow the path it maps out for you.'[122]

Luria explicitly linked the child and the primitive and linked individual and cultural development. He had no interest in glorifying a period of imagined lost innocence, however. In his work with both children and

[118] Benjamin, 'Old Forgotten Children's Books', pp. 406–413, p. 407.

[119] Jacqueline Rose, *The Case of Peter Pan or The Impossibility of Children's Literature* (Philadelphia, PA: University of Pennsylvania Press, 1984), p. 4.

[120] Rose, *Peter Pan*, p. 138.

[121] Rose, *Peter Pan*, p. 8.

[122] Jean Jacques Rousseau, *Émile or on Education*, trans. by Allan Bloom (New York, NY: Basic Books, 1979), p. 47.

Central Asian people, he was intent on tracing a progression from 'lower' to 'higher' modes of thought:

> Each subsequent stage supersedes and supplants the previous one; only after he has gone through the stages of inventing his own devices and has learned the cultural systems evolved over the centuries does a child arrive at a stage of development that is characteristic of advanced, civilized man.[123]

Just as the capacity to count and use language developed in civilisation, so the child must develop these skills as an individual. As he observed in a discussion of experiments with children learning to write: 'Lines and scribbles are replaced by figures and pictures, and these give way to signs. In this sequence of events lies the entire path of development of writing in both the history of nations and the development of the child.'[124] This transition to the historical was imagined as an emancipatory journey away from nature. Modern technology, social institutions and collective ways of living all contributed to creating a more advanced human being, capable of intervening in the world.

After the October Revolution, dominant representations of childhood as a lost idyll were displaced by narratives of hardship and endurance. Leon Trotsky's *My Life* opens with a caustic attack on bourgeois understandings of childhood:

> Childhood is looked upon as the happiest time of life. Is that always true? No, only a few have a happy childhood. The idealisation of childhood originated in the old literature of the privileged. A secure, affluent, and unclouded childhood, spent in a home of inherited wealth and culture, a childhood of affection and play, brings back to one memories of a sunny meadow at the beginning of the road of life. The grandees of literature, or the plebeians who glorify the grandees, have canonised this purely aristocratic view of childhood. But the majority of the people, if it looks back at all, sees, on the contrary, a childhood of darkness, hunger and dependence. Life strikes the weak—and who is weaker than a child?[125]

[123] Luria, 'Experimental Psychology…', p. 95.

[124] Luria, 'The Development of Writing in the Child', *Selected Writings*, pp. 145–194. First published in *Voprosy marksistkoi pedagogikii* [*Problems of Marxist Pedagogy*] (Moscow: Academy of Communist Upbringing, 1929), pp. 143–176.

[125] Leon Trotsky, *My Life: An Attempt at an Autobiography* (Middlesex: Penguin, 1970), p. 1.

Maxim Gorky's *My Childhood* (1913), set in the years immediately preceding the revolution, is a tale of torment and perseverance that became the archetypal narrative of the early years of life in the Soviet Union. In the 1920s and 1930s it became the dominant model for Soviet narratives of childhood, read and re-read across the country.[126] The story opens with the child confronting his father's death. Little Gorky peers into his father's grave and watches in horror and confusion as some little frogs hopping about on the coffin are suffocated beneath the soil. Set in a dull quagmire of murkiness, mist and mud, Gorky describes his childhood as a period of fear, anxiety and suffering: a 'dreary life … full of violence', a 'close-knit, suffocating little world of pain and suffering'.[127] Gorky's depiction of his early life represented a stark break with the traditions of Russian literature epitomised by Tolstoy's *Childhood*, which conformed to the Rousseauian model, portraying childhood, through a misty haze of nostalgia and sentimentality, as a period of purity, virtue and harmony with nature.[128]

According to Denise Riley: 'The figure of the child remains an archetype of a biological-to-social knot … as the exemplar of pure nature, pure biology, capable of arriving at full humanity only through successive forays into a world it stands outside.'[129] Riley contrasts the 'unmuddied' child which development psychology treats as a 'pure object of study' with the 'denser, harder, impenetrably complex' adult 'hopelessly riddled with the world's determinations'.[130] Although Luria does discuss the newborn infant as a universal figure situated at the beginning of human life, he does not exalt this natural starting point. Furthermore, the period of natural passivity he describes is short-lived. The world soon intervenes. Despite Luria's insistence that the external, cultural world would ideally have a positive organising impact on the development of the child, he was forced to contend with a chaotic society. Soviet children were not pure and 'unmuddied'; many were already 'dirtied' by their experiences. The realities of the Soviet cultural world did not correspond to the pedagogical methods and tools implemented in the controlled environments of Luria's laboratories. Luria and his team of researchers confronted this when they

[126] See Andrew Baruch Wachtel, *The Battle for Childhood: Creation of a Russian Myth* (Stanford, CA: Stanford University Press, 1990), pp. 149–150.

[127] Maxim Gorky, *My Childhood*, trans. by Ronald Wilks (Penguin: London, 1966), p. 25.

[128] See Baruch Wachtel, pp. 131–152.

[129] Riley, p. 27, p. 33.

[130] Riley, p. 26.

ventured on to the streets of Moscow to conduct experiments with homeless children.

As part of his work at the Academy of Communist Upbringing, Luria undertook comparative studies of children from different social and class backgrounds. *Speech and Intellect among Rural, Urban and Homeless Children* (1930) was a comparative study of the urban city school child, the rural child and the 'urchin' child, edited by Luria. Such comparative studies were a common mode of pedological investigation, setting out to distinguish between nature and nurture.[131] Indeed, these experiments seem to address questions that a critical article published in *Pedology* in 1932 accused Luria and Vygotsky of ignoring:

> From the standpoint of pedology, what is essential and specific to child development is that the child grows and develops as a social type … the contradictions between the biological and the social must be the fundamental and guiding methodological principle in our investigation of this development. … The child's class status dictates a specific type of upbringing under the conditions of class struggle at a given historical stage and in a specific, concrete, social-class environment.[132]

Unlike the laboratory experiments that characterised Luria's research on writing and play and his experiments with twins, the protocols discussed in Luria's contribution 'A Child's Speech Responses and the Social Environment' set out to analyse how children's thought processes were shaped by their cultural experiences in the contexts of their everyday lives.

[131] See, for example, Semunova and Bolunova, 'Rech' sovremennogo russkogo prolearskogo i krest'ianskogo rebionka doshkol'nogo vozrata' [Speech of the Contemporary Russian Proletarian and Peasant Child at Preschool Age], *Pedologiia*, 2 (1928), 47–60, Syrkin, 'Materialy k voprosu o fizicheskom razvitii proletarskogo i krest'ianskogo rebenka' [Materials on the Question of the Physical Differences between Proletarian and Peasant Children], *Pedologiia*, 1–2 (1929), 133–138, Vilenkina, 'Sravnitel'nyi analiz social'no bytovyh uslovii moskovskih detei raznyh social'nykh grupp' ['A Comparative Analysis of the Social and Living Conditions of Moscow Children of Different Social Groups], *Pedologiia*, 1–2 (1929), 157–166, E. Netchaeva, 'Vilyanie sredy na trudovuyu napravlyennost' podrostkov' [The Influence of the Surroundings upon the Industrial Capacity of Adolescents], *Pedologiia*, 1 (1930), 43–53.

[132] M.P. Feofanov, 'The Theory of Cultural Development in Pedology as an Eclectic Conception with Basically Idealistic Roots', *Journal of Russian and East European Psychology*, 38, 6 (2000), 12–30, pp. 15–16. Originally published as 'Teoriia kul'turnogo razviita v pedologii kak elektricheskaia konseptsiia, imeiushchaia vosnovnom idealisticheskie korni', *Pedologiia*, 1–2 (1932), 221–234.

Luria declared 'it is absolutely meaningless to study children divorced from the environmental factors that shape their mental makeup'.[133] Here he was explicit that 'concrete sociohistorical and cultural conditions'[134] transform behaviour, stating that 'child-subjects should not be isolated from their general historical circumstances'.[135] But his insistence that individuals from different social groups were all of equal intelligence was undercut by his conviction that those social groups themselves were not all at the same stage of development.

In an essay on the relationship between psychoanalysis and history, Dominick LaCapra contends that it is inadequate for the historian of psychoanalysis to accept 'the *analogy* between ontogeny and phylogeny or individual and society'. Instead, he declares that 'one must actively recognise that the analogy itself conceals the most basic interaction of psychoanalysis and sociocultural processes involving social individuals'.[136] Similarly, the macrological historical framework that informed Luria's work did not always map neatly onto his understanding of history as an individual's immediate environment. LaCapra notes in his discussion of Freud that 'history in the ordinary sense often gets lost in the shuffle between ontogeny and phylogeny',[137] and the same is also true in many of Luria's writings on the relationship between child development and the development of civilisation. Except that sometimes history 'in the ordinary sense' sometimes intruded anyway, as was the case with this set of experiments.

Abandoned children (*besprizornye* or *besprizorniki*) were a prominent feature of Soviet urban life. The bourgeois nuclear family was being challenged for ideological reasons but many children were separated from their parents by historical circumstances. Bands of urchins roamed the country often resorting to begging, stealing and sex work to survive. Although homeless children existed in Russia prior to the October

[133] Luria, 'A Child's Speech Responses and the Social Environment' in *Soviet Developmental Psychology*, ed. by Michael Cole (New York, NY: M. E. Sharpe, 1977), pp. 32–64, p. 35, p. 60. Excerpted from the edited collection *Speech and Intellect among Rural, Urban and Homeless Children* [*Rech' i intellekt derevenskogo, gorodskogo i besprizornogo rebenka*] (Moscow: Gosizdat RSFSR, 1930).

[134] Luria, 'A Child's Speech…', p. 32.

[135] Luria, 'A Child's Speech…', p. 64.

[136] Dominick LaCapra, 'History and Psychoanalysis', *Critical Inquiry*, 13, 2 (1987), 222–251, p. 242.

[137] LaCapra, p. 222.

Revolution, in the wake of the violence, famine and epidemics that accompanied the First World War and the Civil War their numbers greatly increased. The Bolshevik government was forced to contend with at least seven million *besprizorniki* on the streets of Soviet cities.[138] Initially, Soviet authorities blamed the phenomenon on the tumultuous Civil War years, framing abandonment as a problem of the past.[139] Networks of children's homes were originally planned to provide child-rearing facilities for working parents, but in the face of mass homelessness tended to function as shelters. The number of children living on the streets diminished in the 1920s, but as the decade continued it became evident that abandonment persisted, forcing the state to address a far more uncomfortable reality: *besprizorniki* were also a phenomenon of the present.[140]

By the early 1930s, the continued existence of vast numbers of homeless youths, many driven to the cities in the wake of collectivisation, had become an embarrassment to the Soviet government who began to deny their existence.[141] Observers unsympathetic to the Soviet cause pointed to these children as embodied evidence of the revolution's failings. In 1931 the former Socialist Revolutionary Party member Vladimir Zenzinov clearly read the *besprizorniki* as products of history and therefore as a damning indictment of Stalin's regime:

> Covered with dust and dirt, with soot and oil, mud and petrol … clothed in rags … thrown up by the waves of misery and social disorder which have washed them into the towns and villages and railways—The deserted children! The deserted children![142]

[138] Nadezhda Krupskaya gave this figure in 1923 and the same figure was repeated in the 1927 edition of the *Great Soviet Encyclopedia*. See Margaret K. Stolee, 'Homeless Children in the USSR, 1917–1957', *Soviet Studies*, 40, 1 (1988), 64–83, p. 69.

[139] A. Zalkind and M. Epshtein, 'Besprizornost'', *Bol'shaia Sovetskaia Entsiklopediia*, tom 5 (Moscow: Aktsioneroe Obshchestvo Sovetskaia Entsiklopediia, 1929), p. 786.

[140] On shifting attitudes to abandoned children in the Soviet Union, see Alan Ball, *And Now My Soul is Hardened: Abandoned Children in Soviet Russia, 1918–1930* (Berkeley, CA: University of California Press, 1994), Juliane Fürst, 'Between Salvation and Liquidation: Homeless and Vagrant Children and the Reconstruction of Soviet Society', *Slavonic and East European Review*, 86, 2 (2008), 231–258 and Marina Goloviznina, 'Politika sotsial'nogo kontrolia prestupnosti nesovershennoletnikh v SSSR (1917-konets 1980-khgg.)', *Zhurnal issledovanii sotsial'noi politiki*, 3, 2 (2005), 223–240.

[141] Ball, *And Now…*, pp. 193–195.

[142] Vladimir Zenzinov, *Deserted: The Story of the Children Abandoned in Soviet Russia* (London: H Joseph, 1931), p. 129.

The glittering, regimented army of dedicated pioneers that marched across Soviet propaganda was haunted by this hungry, foul-smelling horde of young people.

The contradictions of revolutionary childhood were magnified in the figure of the abandoned child. Immediately following the revolution, political idealists had proposed to transform this swathe of children into the new revolutionary generation, accustomed to simple, communal living and untarnished by family life. Zalkind proclaimed confidently that these children were '*closer* to the collective comradely ethic than the "normal" morality of self-love that children from a bourgeois environment have'.[143] But educational institutions focused on rehabilitation and re-education, which proved difficult in practice. Many of those who found their way into the oversubscribed state-run institutions were deeply traumatised by their street lives, others simply ran away. These young people had grown up isolated from social norms: they swore, they smoked, they refused to sit still and they often spoke in a street argot that was incomprehensible to outsiders. Like the *ichkari* women discussed in the previous chapter, these children occupied a liminal social position. This initially marked them out as potential allies in the revolutionary struggle who posed a threat to the bourgeois structures of the past, but they soon figured as a threat to the orderly new social structures being erected in the present.

Langston Hughes, after visiting the Soviet Union in the early 1930s, remarked upon the contradictory qualities of the 'pitiful and half-savage children of upheaval' many of whom resisted the state's attempts to place them on the 'path to progress', preferring their 'precarious freedom' of exciting peregrinations, mischief and petty crime, to the relative 'comfort and regularity' provided by state institutions, declaring: 'These determined little hooligans were making a last stand for freedom'. In Hughes's account the 'wandering delinquents' are presented as cheeky, inventive and charming, a positive reminder of the persistence of irregularity and spontaneity in an increasingly orderly and austere Communist context.[144] Hughes perceived that the children represented a potential challenge to the already stultifying social order, but this was not the line taken by Party elite. On the eighth anniversary of the October Revolution in 1925 Soviet

[143] Zalkind, 'Psikhopaty li besprizonye?' *Vestnik prosveshcheniia*, 9 (1924), p. 136 cited in Gorsuch, p. 160.

[144] Langston Hughes, *I Wonder as I Wander: An Autobiographical Journey* (New York, NY: Thunder's Mouth Press, 1986), pp. 153–154.

President Mikhail Kalinin had declared it necessary to 'save' the *besprizorniki*: 'The situation here threatens grave dangers for the future if we are not able to eradicate promptly in youths the bad habits that a vagrant life imparts to them.'[145] The *besprizorniki* threatened to shatter the 'good habits' espoused by the nascent Communist state.

Besprizorniki often lived on rubbish dumps or coal heaps. Many survived on scraps of discarded food. Alan M. Ball vividly describes homeless children nibbling at fish heads, chewing on watermelon rinds or licking crumbs from plates. They were likened to flocks of sparrows, ravenous locusts or parasites.[146] Konstantin Paustovsky described encountering a group of street children in dirty, over-sized jackets in whose pockets they kept all of their belongings—'bits of broken combs, knives, cigarettes, crusts of bread, matches, greasy cards and bits of dirty bandages'—and beneath which could be glimpsed their 'frozen, dirty, greenish-yellow bodies, covered with bleeding streaks'.[147] These external attributes were taken to reflect the children's messy and unsavoury inner lives. These filthy children were not pristine new babies, lumps of clay ready to be moulded. They were not raw material but had already been battered by the ravages of experience. Soviet children's institutions were initially imagined as the forges of the new nation, but the reality was closer to a salvage operation.[148]

In his experiments comparing children from three social groups, including *besprizorniki*, Luria set out to show that speech development was tied to a child's environment:

> A person's class and his particular social experience fill his mind with a quite specific content, and the study of this content not only is of considerable interest for the infant science of psychology of classes but is also of indisputable psychological interest.[149]

Speech, which he viewed as 'the product of specific historical circumstances', acted like a mirror of a child's social surroundings, in that 'it

[145] *Pravda*, 1925, no. 255 (November 7), p. 2 cited in Ball, *And Now...*, p. 193.

[146] Ball, *And Now...*, p. 45.

[147] Konstantin Paustovsky, *The Restless Years* (London: Harvill, 1974), p. 46.

[148] For a discussion of how metaphors of moulding, forging and building were applied to the *besprizorniki*, see Marina Balina, 'It's Grand to be an Orphan!': Crafting Happy Citizens in Soviet Children's Literature of the 1920' in *Petrified Utopia: Happiness Soviet Style*, ed. by Marina Balina and Evgeny Dobrenko (London: Anthem Press, 2009), pp. 99–114.

[149] Luria, 'A Child's Speech...', p. 35.

faithfully reflects the distinctive features of the environment from which he sprang'.[150] Luria thus investigated children's responses to verbal stimuli—presenting them with single words and asking them to respond by association, recording responses and reaction times in a similar method to that used in his experiments with criminals that form the basis of *The Nature of Human Conflicts* (1932), though with less emphasis on the physiological component.[151] Environment was assumed to have an impact on both the form and content of thought, determining 'a vast range of characteristics inherent in the basic mechanisms underlying a person's reactions and, ultimately, the overall pattern of the reactions of the social group to which the individual belongs'.[152] It was not only concluded that rural children knew more about farm animals than urban children, for example, but that the slower responses of the former were related to the slow pace and stability of their environment (he does not give any details about the rural area he is discussing and speaks in general terms, but it is worth bearing in mind that many rural areas of the Soviet Union were far from stable in the years of the First Five Year Plan).

Luria's conclusions were based on the extent to which children from particular groups responded with shared or divergent responses to words, which allowed him to assess the relative 'uniformity of collective experience'.[153] Luria noted a contrast between 'accidental, individualised memories', more common among the street children, versus 'routine associations common to the entire group'.[154] He approvingly recorded that educated urban children were able to provide a range of associations to words like 'rake' or 'cart', which would have been familiar to them only indirectly through education, rather than through their direct experiences:

> Obviously school studies and the influences of city life expose the child to experiences in all spheres of life, and there is practically nothing that is totally foreign to him. The material to which he is exposed in his schoolwork makes

[150] Luria, 'A Child's Speech…', p. 33.

[151] Luria extended his 'combined motor method' in experiments with children conducted in the 1950s with S.V. Yakoleva and O.K. Tikhomirov, which asked a child of 2.5 years to squeeze a bulb or cry out in response to a signal. See A.R. Luria, 'A.R. Luria' in *A History of Psychology in Autobiography*, vol. 6, ed. by Gardner Lindzey (Cambridge, MA: Harvard University Press, 1974), pp. 253–292, pp. 272–273.

[152] Luria, 'A Child's Speech…', p. 36.

[153] Luria, 'A Child's Speech…', p. 49.

[154] Luria, 'A Child's Speech…', p. 44.

> nature and society just as accessible to him as the direct impressions he receives from his environment. For the city child the world of immediate impressions begins to give way to more mediated forms of experience, to world of facts assimilated through communication, books, and cultural influences.[155]

The homogeneity of the rural children's outlook was attributed to their relatively stable and unchanging experiences, whereas the urban school children were shown to be developing a more abstract notion of shared identity. He compared the responses of rural and urban children to the word 'Union' claiming that the former tended to understand the word 'in the sense of a cooperative union, a shop, where they could buy calico, boots etc., whereas the urban children, without exception, understood union to refer to a social or political organisation';[156] the former set of responses was more concrete, the latter more abstract. When asked to respond to the word 'truth' he deemed rural children's responses—which included 'Dove', 'Crookedness', 'Grandpa', 'Cat', 'Old lady'[157]—'inappropriate in the vast majority of cases', whereas the responses of urban children 'clearly demonstrated that they had an adequate grasp of the word's meaning'.[158]

According to Luria, homeless children experienced the world 'kaleidoscopically' due to the instability of their daily lives. With the *besprizorniki* 'a collective experience common to the entire group is nonexistent' and they were also unfamiliar with 'areas of experience that are most commonly part of the collective experience of the normally developing child'.[159] Rural children gave very similar responses to the word 'house', with the responses 'hut' and 'barn' predominating, whereas street children gave a huge array of responses to the word indicating their lack of a stable home environment. One child simply responded: 'I don't know'.[160] Luria also compared rural children's and street children's responses to the word 'kettle', which he argued betrayed differences in experience rather than in intellect. The former associated the word solely with eating and food

[155] Luria, 'A Child's Speech…', p. 51.

[156] Luria, 'A Child's Speech…', p. 54. Urban children's responses included 'Workers', 'Councils', 'USSR' and 'Trade unions', p. 55.

[157] Luria, 'A Child's Speech…', p. 55.

[158] Luria, 'A Child's Speech…', p. 54.

[159] Luria, 'A Child's Speech…', p. 57.

[160] Luria, *Speech and Intellect…*', p. 58.

preparation, whereas the latter's responses gestured towards other situations in which they might have encountered a kettle, including the 'street kettles around which the homeless and abandoned gather to keep warm' and boilers or kettles found on ships and trains.[161]

The *besprizorniki* emerged as the slowest respondents to the word associations, though Luria conceded this was not through any intellectual deficit or fault of their own, acknowledging that these children were 'different' rather than 'more backward' than the other groups with whom he conducted these experiments.[162] Luria's experiments revealed the abjection and social ostracisation experienced by these children who had 'grown up without any direct exposure to the influence of stable, familiar, more or less unchanging circumstances'.[163] The homeless children were unfamiliar with formal reasoning and therefore responded slowly to the psychologists' questions:

> Anything even resembling a problem divorced from real life is a novelty to him. His linguistic experience has nothing to do with abstract logical operations of the kind cultivated in school. This is why the urchin, who is quick and deft at coping with the situations of everyday life, becomes quickly disoriented and proceeds slowly in the contrived situations of an experiment, in which, moreover, we employed abstract linguistic (associative) operations.[164]

Furthermore, the homeless children were—like many of the uneducated peasants Luria encountered in Uzbekistan, and presumably also for experiential reasons—mistrustful of the adults conducting the experiments: 'when he is suddenly exposed to the unnatural conditions of a test, he displays a marked emotional tension.'[165] Luria speculated that the hostility and reluctance to participate he encountered among this social group might have been connected to the children's desire to conceal 'any number of questionable deeds in their past',[166] but sympathetically noted that the 'affective traces' or 'affective scars' of previous experiences also played a part in slowing down their responses to the experiments: 'The urchin has had to wage a bitter struggle with life, and that struggle has left its specific

[161] Luria, *Speech and Intellect…*, p. 60.
[162] Luria, 'A Child's Speech…', p. 38.
[163] Luria, 'A Child's Speech…', p. 58.
[164] Luria, 'A Child's Speech…', pp. 38–39.
[165] Luria, 'A Child's Speech…', p. 39.
[166] Luria, 'A Child's Speech…', p. 39.

emotional traces on him.'[167] Having declared speech a 'reflection' of a child's social environment (again borrowing Lenin's concept, discussed in the previous chapter), Luria was forced to acknowledge in his discussion of his encounters with *besprizorniki* that the emotional states associated with difficult or traumatic childhood experiences could act as a complicating factor, like condensation clouding the surface of a mirror.

The homeless children may have scored low on standardised intelligence tests like those devised by Alfred Binet, but Luria insisted that 'in potential they are fully normal; and under the right conditions, they could develop into complete, highly productive, human beings'.[168] He cited research undertaken by Mirenova demonstrating that 'instead of defective intelligence, [the *besprizorniki*] have abilities that have evolved along unique and independent lines'.[169] Luria may have rejected the inherent 'backwardness' of individual children from different social groups but he nonetheless retained a hierarchical developmental model for evaluating differences between those social groups, thus retaining a value-laden framework for comparing people based on suppositions regarding historical rather than individual development: 'The developmental differences we noted in our children were superimposed, so to speak, on underlying class differences of a more basic nature.'[170] Luria emphasised the inability of street children to comprehend both abstract concepts and concrete objects beyond their direct experiences as these 'children had not yet been exposed to structured collective experience'.[171] Although he did not consider the internal rules, languages and norms that structured the lives of Soviet street children[172]—a vagrant child may not have been able to describe the interior of a house or to define a Trade Union or *kolkhoz*, but the words that peppered homeless children's vocabulary—including slang terms for cigarette butts, cocaine, pickpockets, policemen and prisons—

[167] Luria, 'A Child's Speech…', p. 39.

[168] Luria, 'Speech and Intellect…', p. 41. Luria also criticises psychometric tests in 'Variability in Mental Development', pp. 73–74. On the issue of testing more broadly, see Andy Byford, 'The Mental Test as a Boundary Object in Early-20th-Century Russian Child Science', *History of the Human Sciences*, 27, 4 (2014), 22–58.

[169] Luria, 'Speech and Intellect…', p. 40.

[170] Luria, 'Speech and Intellect…', p. 44.

[171] Luria, 'Speech and Intellect…', p. 50.

[172] On these internal rules, see Zenzinov, p. 127.

would presumably have been equally as incomprehensible to a rural or urban schoolchild.[173]

It was this unique language that Benjamin declared pedagogues needed to penetrate in order to communicate with the 'infantile masses': 'To get through to them at all, to he heard one has to relate as directly and clearly as possible the catchwords of the street itself.'[174] These words were key to understanding their 'collective life' whose structure was, Benjamin insisted, inherently political: 'Politics, in the organisation of crowds of such children, is not tendentious, but as natural a subject, as obvious a visual aid, as the toyshop or dollhouse for middle-class children.'[175] Benjamin suggested that the visitor to Moscow should attempt to see the city from the viewpoint of its street children: 'One ought to know Moscow as such beggar children know it'.[176] Though this might sound at risk of romanticising street homelessness, Benjamin's work suggests that in order to finally rid the world's streets of the misery embodied in figures like the street child (or the ragpicker who wanders through *The Arcades Project*), history should be approached from their perspective, as a heap of rubble from which it might still be possible to wrest some sparkling shards with which to build anew. For Luria such an approach would have necessitated looking down from the gleaming progressive road his work attempted to map out to examine the potholes and litter on the streets beneath his feet.

Conclusion

On 5 July 1936 the front page of *Pravda* was dominated by the news of the CPSU(b) Central Committee decree under the headline: 'On Pedological Perversions in the Narkompros System'.[177] In the wake of the decree an article criticising Vygotsky's work declared 'pedology is not a science that has been promoted by life and human practice, but something contrived'.[178] Pedologists were attacked for emphasising immediate envi-

[173] Ball lists some of the terms specific to the dialect of *besprizornye*, p. 38.

[174] Benjamin, 'Moscow', p. 27.

[175] Benjamin, 'Moscow', p. 27.

[176] Benjamin, 'Moscow', p. 28.

[177] 'O Pedologicheskikh Izvrashcheniiakh v Sisteme Narkomprosov' [On the Pedological Perversions of the Narkompros System], *Pravda*, July 5, 1936, p. 1.

[178] A.V. Kozyrev and P.A. Turko, 'Professor L.S. Vygotsky's "Pedagogical School"', *Journal of Russian and East European Psychology*, 38, 6 (2000), 59–74, p. 69. Originally published as '"Pedologicheskaia shokla" Prof L.S. Vygotskogo', *Vysshaia Shkola*, 2 (1936), 44–57.

ronmental factors and biological essences over psychological propensities, for emphasising passivity over activity. Pedological publications focusing on Soviet children growing up in different social milieus, employing methodologies similar to Luria's study of the *besprizorniki*, were condemned. Like Luria's experiments in Central Asia, they revealed how far the Soviet Union was from constructing its ideal model citizens and implied that the proletariat was impervious to change. It was alleged that the reports and the tests pedologists espoused tended towards description rather than intervention, again echoing the arguments deployed to attack Luria's research in Uzbekistan. Vygotsky was accused of overemphasising biological factors and of espousing the 'anti-Marxist, antiscientific "theory" of equilibrium, the "theory" of man's adaptation to the environment ... i.e. to subordinate the child to his environment'.[179]

Despite their revolutionary zeal, pedologists were forced to recant and furiously attempted to make amends: 'The fact that subjectively we considered our theory revolutionary has no relevance. By excluding the purposive [*tselepolagayushchaia*] activity of man, his interference in the course of events, we automatically excluded also the necessity of educative work.'[180] Pedology was declared a failure as it was said to describe human beings helplessly responding to their environment or acting in accordance with hereditary factors, rather than fostering dedicated revolutionaries intent on redirecting the course of history. Zalkind is alleged to have dropped dead upon hearing the news of the discipline's official denunciation.[181] Vygotsky's work was blacklisted until the end of the Stalin era.[182] Luria's experiments with children were terminated; he lost his position at

[179] E.I. Rudneva, 'Vygotsky's Pedological Distortions', *Journal of Russian and East European Psychology*, 38, 6 (2000), 75–94, p. 91. Originally published in *Pedologicheskie izvrashcheniia Vygotskogo* (Moscow: Gosudarstvennoe uchebno-pedagogicheskoe izdatel'stvo, 1937), pp. 3–32.

[180] A.S. Zaluzhnyi, *Lzhenauka pedologiia v 'trudakh' Zalkinda* (Moscow, 1937), pp. 20–21, cited in Sheila Fitzpatrick, *Education and Social Mobility in the Soviet Union, 1921–1934* (Cambridge: Cambridge University Press, 1979), pp. 229–230. This echoes denunciations of Vygotsky: 'Vygotsky does not understand the Marxist-Leninist theory of the environment; he disregards the role of man in transforming the environment', Rudneva, p. 93.

[181] Etkind, *Eros of the Impossible*, p. 277.

[182] The precise implications of the decree and the status of Vygotsky's work during the Stalin period are examined in Jennifer Fraser and Anton Yasnitsky, 'Deconstructing Vygotsky's Victimization Narrative: A Re-examination of the "Stalinist Suppression" of Vygotskian Theory', *History of the Human Sciences*, 28 (2015), 128–153.

the Medico-Genetic Institute and turned to focus on neurological research at the Institute of Neurosurgery (later named after N.N. Burdenko). It is unclear what happened to the numerous sets of twins under observation in Moscow.[183] By the time of the decree, Stalin had consolidated his position and many of the radical assumptions about childhood and family life that had characterised the post-revolutionary period were overturned.[184]

Luria's research with children emerged from a period of multivalent approaches to child psychology and education. Anton Makarenko's work dominated Soviet pedagogy in the wake of the decree on pedology, advocating strict, militaristic authority: 'austere rhythm, iron discipline and distinguished bearing'.[185] Although even the most utopian visions of childhood developed in the aftermath of revolution tended to imagine the transition to adulthood as a transition from chaos to order, paralleling the Leninist narrative of the movement of history towards communism, Anna Louise Strong's depiction of the experimental John Reed Children's Colony published in 1925, for example, concluded by contrasting the 'weeds, broken fences, disorder' of the old landscape with the 'acres of ploughed fields' across which a young Communist hero advances in his tractor.[186]

Disrupted, disorganised, impulsive and spontaneous—the newly born child in Luria's account is a chaos who perceives the world as an undifferentiated confusion of 'noises and splotches'.[187] He and Vygotsky wrote that the young child performs 'an imbroglio of actions ... [a] rich and

[183] In the Preface to the first English edition of *Speech and the Development of Mental Processes in the Child* in 1958, Luria wrote he had 'long lost sight of the two twins, Yura and Liosha', p. 16.

[184] In her wide-ranging discussion of Russian childhood, Catriona Kelly considers 1935 as a watershed year in the history of Soviet childhood. She considers the period from the revolution until 1935 as a discrete epoch characterised by competing but often radical ideas about the child. In addition to the decree on pedology, 1935 and 1936 saw major policy reforms that effected children. Legislative changes included lowering the age of criminal responsibility, cracking down on waifdom [*besprizornost'*], an upsurge in propaganda aimed at families and policy shifts in schools that placed an increased emphasis on authority and discipline. See Kelly, *Children's World*, pp. 61–92.

[185] Anton S. Makarenko, *The Road to Life: An Epic in Education* (New York, NY: Oriole Editions, 1951), p. 414.

[186] Anna Louise Strong, *Children of Revolution: Story of the John Reed Children's Colony* (Seattle, WA: Piggott Press, 1925), p. 98.

[187] Luria, 'The Child...', p. 145.

often grotesque mixture of mutually contradictory forms'.[188] The child's 'splintered and peculiar' perception was presented as a mess that would eventually be tidied up.[189] According to Luria, the chaotic inner life of the baby was soon replaced by the order of the cultural world, a process that could be hastened and improved through education. Luria distinguished between chaos and order, irrationality and reason, ignorance and knowledge as successive stages. Even taking this linear understanding of individual development on its own terms, however, it was still troubled by the coexistence of children with adults, their ongoing interactions and interruptions. After all, new 'disorderly' babies continued being born. Luria may have dispensed with the Freudian unconscious, a dangerously prehistoric internal reservoir, but he could not deny the external existence of children who he and Vygotsky described in terms strangely reminiscent of Freud's definition of the unconscious: 'There are no contradictions for the child, he does not see them, contradictory judgments can coexist not excluding one another.'[190] Luria's naturally chaotic newborn infant had a historical counterpart in the figure of the abandoned child; the existence of *besprizorniki* pulled against the supposedly inevitable course of both individual and historical development. In Luria's account of development, the incoherence of the baby could soon be overcome, even as the actual children he encountered in his research did not always corroborate this thesis.

Five years after the decree that banned pedology, Germany invaded the Soviet Union. The Second World War was to have a lasting impact on Luria's work. The pinnacle of development in Luria's work was the 'civilised', educated adult but this was precisely the type of person who ended up on Luria's wards with bullets in their brains. Benjamin perceived that the 'vulgar Marxist' celebration of the progressive mastery of humanity over nature failed to acknowledge the 'retrogression of society' that accompanied it.[191] As Adorno and Horkheimer declared in *The Dialectic of Enlightenment*: 'Civilization leads back to the terrors of nature'.[192] Work with brain-injured soldiers dominated Luria's wartime work; wounded people he was

[188] Luria and Vygotsky, 'Tool…', p. 117.

[189] Luria and Vygotsky, 'Tool…', p. 125.

[190] Luria and Vygotsky, 'The Child…', p. 164.

[191] Benjamin, 'Theses on the Philosophy of History' in *Walter Benjamin: Selected Writings, Vol. 4, 1938–1940*, ed. by Howard Eiland and Michael W. Jennings, trans. by Edmund Jephcott (Cambridge, MA: Belknap Press of Harvard University Press, 2006), pp. 389–400, p. 393.

[192] Theodor Adorno and Max Horkheimer, *The Dialectic of Enlightenment: Philosophical Fragments*, ed. by Gunzelin Schmid Norr, trans. by Edmund Jephcott (Stanford, CA: Stanford University Press, 2002), p. 89.

dedicated to healing. In these cases, however, incoherence could not be overcome. The process of healing proved partial but not futile, forcing Luria to confront the persistence of the shattered in the midst of habitual life, without giving up hope in the possibility for both individual and social transformation.

CHAPTER 5

The Aphasic

In Leo Tolstoy's *War and Peace* (1865–1868) the character Andrei Bolkonsky is wounded in the Battle of Austerlitz and awakens on the battlefield in a state of confusion:

> On the Pratzen heights, at the spot where he had fallen with the flagstaff in his hand, lay Prince Andrei Bolkonsky, losing blood and, without realising it, moaning a soft, plaintive moan like a child.
>
> Toward evening his complaining ceased and he became quite still. He did not know how long his unconsciousness lasted. Suddenly he felt again that he was alive and suffering from a burning, lacerating pain in his head.
>
> "Where is the lofty sky I saw today and had never seen before?" was his first thought. "And this agony I did not know either", he thought. "Yes, I knew nothing, nothing till now. But where am I?"[1]

A century after the novel's publication Alexander Luria wrote to his friend Jerome Bruner from the estate of the aristocrat Sergei Volkonsky upon whom Tolstoy based Andrei Bolkonsky. He declared he was at work on his own 'neurological novel'.[2] Luria's book would also describe a man wounded in battle who awakens from unconsciousness like a 'newborn

[1] Leo Tolstoy, *War and Peace*, trans. by Rosemary Edmunds (London: Penguin, 1978), p. 338.

[2] Luria to Bruner, December 20, 1969, Cambridge, MA, Harvard University Archives (HUA), Jerome Bruner Papers, General Correspondence 1975–1977, HUG 4242.5, Box 88.

H. Proctor, *Psychologies in Revolution*, Mental Health in Historical Perspective, https://doi.org/10.1007/978-3-030-35028-4_5

creature' to a life he had not previously known.[3] Luria described the case history as 'a little book about a great tragedy—a man with a bullet in his brain who lost his world, his past and his future'.[4] *The Man with a Shattered World: The History of a Brain Wound* (1972) focuses on Zasetsky, a former Red Army soldier who fought in the Second World War, surviving a bullet wound to the head that permanently damaged his cerebral functioning. Like Tolstoy's epic novel, Luria's little book also offers a comment on its epoch, a meditation on war and peace and a sombre reflection on Russian history.

This chapter situates Luria's history of a brain wound in the shattered world that produced it; the bullet that penetrated Zasetsky's brain and the burnt tissue it left behind is treated as a metonym for the impact of the Second World War on Soviet history. Luria composed the text in a self-consciously literary style. As such, this chapter situates *The Man with a Shattered World* in the context of Soviet literary debates. However, the non-fictional status of the book—the material conditions it arose from, the neurological condition it described and the therapeutic process Zasetsky engaged in—forced Luria to deviate from existing narrative models. As with his experiments with 'primitives' and children, in his work with brain-injured people Luria hoped to facilitate and describe a progressive movement from chaos to order, a movement that paralleled his understanding of civilisational development. Unlike in the developmental paths traced in the preceding chapters of this book, where Luria hoped order would eventually triumph over chaos, *The Man with a Shattered World*, and the patient it described, was incapable of reaching a final moment of synthesis. Although the wounded subject Luria described lost his connection to the past and the present, he ultimately retained hope for the future. Luria's text attests to the lasting damage of his patient's injury—it describes the interminability of his struggle to heal, but still resists sinking into apathy.

[3] Alexander Luria, *The Man with a Shattered World: The History of a Brain Wound*, trans. by Lynn Solotaroff (Cambridge, MA: Harvard University Press, 2002), p. 10. First published in Russian as *Poteriannyi i vozrashchennyi mir: Istoriia odnogo raneniia* [*A World Lost and Re-gained: The Story of an Injury*] (Moscow: Izdatel'stvo MGU, 1971).

[4] Luria to Bruner, December 20, 1969, HUA, Jerome Bruner Papers, General Correspondence 1975–1977, HUG 4242.5, Box 88.

New Wounded

In *The New Wounded* (2012), Catherine Malabou asks what mode of expression might adequately capture the experience of her eponymous injured figures. For Malabou, the 'new wounded' are people who have experienced a severe neuronal disturbance. Her definition encompasses post-traumatic shock, degenerative disease and brain injury, and is metaphorically extended to include any experience that dramatically interrupts the course of a person's life. The condition, she insists, takes a single form: it is characterised by 'coolness, neutrality, absence, and the state of being emotionally "flat"'.[5] The 'new wounded' live in a perennial present, cut-off from past identity and future hopes; their suffering is a peculiar form of pain that 'manifests as an indifference to pain, impassivity, forgetting, the loss of symbolic reference points'.[6] She sees this indifferent subjectivity reflected in the apathy of the contemporary moment.

Malabou approaches the question of representing the subjectivities of the 'new wounded' through a consideration of case histories dealing with people with serious neurological defects. She discusses Luria's case history *The Man with a Shattered World* alongside works by Oliver Sacks, criticising both authors for what she sees as a false imposition of affect onto their subjects, for their attempts to 'weave the patient's coolness, indifference, and the disintegration of emotion into a narrative intrigue that must not be disaffected itself'.[7] The paradox she identifies in such an exercise is that by transforming their subjects into exemplary cases, their narratives cannot be as detached as she presumes their patients'

[5] Malabou, *The New Wounded* (New York, NY: Fordham University Press, 2012), p. 53. For an in-depth consideration of the political implications of Malabou's work, see my review essay: 'The Post-Traumatic Condition', *Radical Philosophy*, 177 (2013), 40–43. My interest in Luria and in Malabou's discussions of his work began as an MA essay, an edited version of which was published as Hannah Proctor, 'Neuronal Ideologies: Catherine Malabou in Light of A. R. Luria', *Dandelion: Postgraduate Arts Journal and Research Network*, 2, 1 (2011). Parts of an early draft of this chapter were also included in 'The History of a Brain Wound: Alexander Luria and the Dialectics of Soviet Plasticity' (co-written with Laura Salisbury) in *Plasticity and Pathology: On the Formation of the Neural Subject*, ed. by David Bates and Nima Bassiri (Bronx, NY: Fordham University Press, 2015). The second edition of Malabou's *What Should We Do With Our Brain?* was recently translated from French into Russian and she cites my critical engagements with her work in her preface. See, Katrin Malabu, *Chto nam delat' s nashim mozgom?* (Moscow: VAG Press, 2019).

[6] Malabou, *The Ontology of the Accident: An Essay in Destructive Plasticity*, trans. by Carolyn Shread (London: Polity Press, 2012), p. 18.

[7] Malabou, *Wounded*, p. 55.

subjectivities are. According to Malabou these works impose a redemptive structure on the patients' conditions, rather than displaying a form of psychic damage that she characterises as being fundamentally beyond healing. Malabou's 'new figures of the void' can only be thought in relation to what she calls destructive plasticity, a plasticity 'that does not repair, a plasticity without recompense or scar, one that cuts the thread of life'.[8] Following both cerebral and social 'accidents', an entirely new subject emerges, totally discontinuous from any previous identity. The identity that does emerge is, she says, blank and featureless, devoid of any capacity or will to actively forge a new identity in the future.

Malabou's adherence to a singular definition of post-traumatic subjectivity precludes her from analysing the ways Luria's case history fundamentally differs from Sacks's. One major reason for this is her failure to consider the huge spectrum of responses that might be produced by brain damage. Sacks's best-selling *Man Who Mistook His Wife for a Hat* (1985) focuses on people with right-hemisphere disorders. With injuries to the right hemisphere, the patient not only loses their grasp of reality but is also unaware of this loss.[9] As Jacqueline Rose discusses, Sacks presents a crisis of 'the narrative function itself, which is either missing, frozen in time, or else acted out in a type of wild and meaningless superabundance'.[10] This, Rose is quick to point out, distinguishes Sacks's work from both Freud's and Luria's case histories. Sacks's cases are populated by people who are in a state of constantly renewed crisis; there is no progression or resolution. Sacks describes people who experience time as suspended but lack any awareness or anxiety about its suspension.[11] The characters that populate *The Man Who Mistook His Wife For A Hat* might indeed be said to exhibit some of the qualities Malabou associates with the 'new wounded'. In contrast, Luria's patient Zasetsky did not sustain damage to the frontal cortex. Despite the enormous problems he faced, this enabled

[8] Malabou, *Ontology*, p. 6.

[9] Although Luria died before the publication of *The Man Who Mistook his Wife for a Hat*, he reflected on the distinction between left- and right-hemisphere disorders in a letter to Sacks, noting that right-hemisphere disorders lead to 'a loss of awareness'. Luria to Sacks, August 4, 1975, Oliver Sacks Archives, Oliver Sacks Foundation, New York City.

[10] Jacqueline Rose, 'The "The Man Who Mistook His Wife for a Hat" or "A Wife is like an Umbrella"-Fantasies of the Modern and Postmodern', *Social Text*, 1 (1989), 237–250, p. 238.

[11] See, Oliver Sacks, *The Man who Mistook his Wife for a Hat and Other Clinical Tales* (New York, NY: Summit Books, 1985).

him to retain the capacity to make and carry out plans. Sacks contrasted his personal experience of a broken limb to Zasetsky's brain injury, noting that the 'fragments of function' Luria's patient retained enabled 'systematic reconstruction, joining together the bits which had been sundered', whereas his injury though more specific and confined left behind 'not a trace, not a ghost ... not the faintest or most fleeting residue of function'.[12] The exact cerebral location of Luria's patient's injury also accounted for his acute sense of loss, his painful, conscious awareness of what had befallen him and thus enabled his determination to recover.[13]

Malabou follows Sacks in considering his work as part of a single continuum with Luria's.[14] In an introduction to *The Man with a Shattered World*, Sacks claims Luria's work was 'always and centrally concerned with identity' and suffused with 'warmth, feeling and moral beauty'.[15] In his first letter to Luria in 1973, Sacks emphasised that human universality transcended the geographical and ideological separations between himself and Luria:

> We are so strangely different, so strangely similar! Your letter came to me like a revelation from 'the other side', piercing the 'Iron Curtain', showing that in reality, in humanity there is no 'other side', only foolish barriers which cut us off from each other.[16]

Although elsewhere Malabou provocatively argues that brains should be understood in relation to politics and history, both Sacks and Malabou

[12] Sacks continues to expand on this contrast later in the same letter: 'The Lurian method reconstitutes knowledge, but presupposes an affirmative will-to-be, to have, to become. ... I felt I had no cognitive starting-point for the reconstitution of knowledge, but that there was an intense disbelief and disaffirmation.' Sacks to Luria, June 30, 1975, Oliver Sacks Archives, Oliver Sacks Foundation, New York City. In *A Leg to Stand On*, which Sacks dedicated to Luria and about which they corresponded at length, Sacks notes that most of the examples discussed in Luria's *Higher Cortical Functions in Man* are similarly concerned with the left hemisphere (Oliver Sacks, *A Leg to Stand On* (London: Duckworth, 1984), p. 165). See Luria, *Higher Cortical Functions in Man*, trans. by Basil Haigh (New York, NY: Basic Books 1966). Originally published as *Vysshie korkovye funktsii cheloveka* (Moscow: Izdatel'stvo MGU, 1962).

[13] Luria, *Shattered*, p. 34.

[14] Similar assumptions about Sacks's and Luria's case histories are also made in Anne Harrington, 'The Inner Lives of Disordered Brains', *DANA Foundation*, 1 (2005) and Nikolas Rose and Joelle M. Abi-Rached, *Neuro: The New Brain Sciences and the Management of the Mind* (Princeton, NJ: Princeton University Press, 2013), p. 206.

[15] Sacks in Luria, *Shattered*, p. xvi.

[16] Sacks to Luria, August 2, 1973, Oliver Sacks Archives, Oliver Sacks Foundation, New York City.

tend to downplay the historical and cultural specificity of Luria's case histories (even if Sacks was more attuned to differences between neurological conditions than Malabou). In her discussion of Luria and Sacks Malabou fails to apply her own insight that 'any vision of the brain is necessarily political'.[17] She asserts that the 'new wounded' are emblematic of the political torpor of the neoliberal present, insisting that destructive plasticity 'refuses the promise, belief, symbolic constitution of all resources to come',[18] but operates in the open space left in the wake of the 'collapse of messianic structures'.[19] Destructive plasticity refuses salvation and redemption, placing a prohibition on thinking other possibilities: 'It has nothing to do with the tenacious, incurable desire to transform what has taken place.'[20] Instead, she claims, destructive plasticity 'deploys its work starting from the exhaustion of all possibilities, when all virtuality has left long ago, when the child in the adult is erased, when cohesion is destroyed, family spirit vanished, friendship lost, links dissipated in the ever more intense cold of a barren life'.[21] By giving Luria's case its history back and by analysing the text's peculiar formal qualities, this chapter demonstrates the ways his dialectical account of brain injury combines destruction with creation, woundedness with healing and fragmentation with unity. Despite the severity of the injury it describes, *The Man with a Shattered World* insists on thinking other possibilities and on dialectically transforming what has taken place. By returning to the past, Luria's case history also suggests a way of thinking beyond the desolate neoliberal present Malabou describes.

Wartime Work: Adaptation and Restoration

Luria begins *The Man with a Shattered World* with a brief description of Zasetsky's life before his brain injury: 'In the beginning it was all so simple. His past was much like other people's: life had its problems, but was simple enough, and the future seemed promising.'[22] The narrative voice then switches. Zasetsky's first-person narrative begins by stating: 'Then

[17] Malabou, *What Should We Do With Our Brain?* trans. by Sebastian Rand (New York, NY: Fordham University Press, 2008), p. 52.

[18] Malabou, *Ontology*, p. 88.

[19] Malabou, *Ontology*, p. 88.

[20] Malabou, *Ontology*, p. 89.

[21] Malabou, *Ontology*, pp. 89–90.

[22] Luria, *Shattered*, p. 3.

suddenly it was all over'.[23] The sudden event that brought his previous life abruptly to an end was not his injury, however, but the Soviet declaration of war: 'Early one morning I was headed for the institute, thinking about my future, when suddenly I heard, actually shuddered at, the terrible news: there was a war on with Germany!'[24] The war not only interrupted the present, it also blocked the path to the future. Zasetsky's life story is thus placed in the framework of historical events.

On 22 June 1941, Germany broke its non-aggression pact with the Soviet Union. It was two weeks before Stalin finally addressed his Soviet 'brothers and sisters', mobilising the masses for 'a war of the entire Soviet people' in which an estimated 9 million Soviet soldiers alone (and an estimated 25 million citizens in total) would meet their deaths.[25] Like the October Revolution and Stalinist 'great break' that preceded it, the Second World War was a violently transformative moment in Soviet history (and hence in Zasetsky's life and Luria's career). Amir Weiner challenges the view that the October Revolution and subsequent socio-economic transformations of the 1920s and 1930s should be seen as the exclusive 'formative and enduring moments' in the life of the Soviet polity.[26] The Second World War, he argues, played as crucial a function in reshaping the revolutionary narrative, disrupting and consolidating Soviet identities in new ways, and forging myths that would ultimately outlive the Soviet Union itself. Weiner conceives of the revolution as a constantly unfurling project punctuated by a series of traumatic events that both shaped and were shaped by the ongoing enterprise of building communism:

> Within this chain of cataclysms, the war was universally perceived as the Armageddon of the Revolution, the ultimate clash dreaded yet expected by the first generation to live in socialist society, the event that would either vindicate or bring down the system.[27]

[23] Luria, *Shattered*, p. 6.

[24] Luria, *Shattered*, p. 6.

[25] Richard Overy, *Russia's War: A History of the Soviet War Effort: 1941–1945* (London: Penguin, 1997), p. 79.

[26] Amir Weiner, *Making Sense of the Second World War and the Fate of the Bolshevik Revolution* (Princeton, NJ: Princeton University Press, 2001), p. 14.

[27] Weiner, p. 17.

The almost inconceivably brutal experience of the war had a lasting impact on Soviet subjectivities and the archetypes, narratives and images that claimed to represent and helped to shape those subjectivities.

Psychologies in Revolution has sought to demonstrate that Luria's career was shaped by pivotal events in Soviet history. As with the major historic events that preceded it, echoes of the official narratives identified by Weiner can be found in Luria's work. In his autobiography he writes:

> World War II was a disaster for all countries, and it was particularly devastating for the Soviet Union. Thousands of towns were destroyed, tens of thousands of people died from hunger alone. Many millions, both civilian and military, were killed. Among the wounded were thousands who suffered brain injury and who required extended painstaking care. The unity of purpose of the Soviet people so clearly felt during the great Revolution and the subsequent years re-emerged in new forms. A sense of common responsibility and common purpose gripped the country. Each of us knew we had an obligation to work together with our countrymen to meet the challenge. We each had to find our place in the struggle.[28]

Luria clearly situated his own medical work within the broader collective war effort. Also, in conformity with Weiner's observations, he explicitly positioned the war in relation to the project of the revolution. The war appears as a devastating event that united the nation around a shared goal.

Following the outbreak of war, Luria briefly joined the volunteer corps before being commissioned to organise a hospital to care for those who required treatment away from the front. In the 1930s, Luria had entered medical school and retrained as a neurologist, graduating from medical school in 1937. During the war, the focus of his work was on people who had survived traumatic brain injuries. Aphasia, a neurological condition resulting in the partial or complete loss of speech, had been the focus of a series of clinical investigations conducted during Luria's time in Kharkiv and became the main focus of his work during the war.[29]

In 1942 Luria set up a 400-bed sanatorium known as Evacuation Hospital No. 3120 in Kisegach near Chelyabinsk in the Southern Urals,

[28] Luria, *Making of Mind*, p. 138.

[29] See the bibliography in G. Braemer and W. Jantzen, 'Bibliographie der Arbeiten von A.R. Lurija' in *Die neruolen Verstickungen des Bewusstseins* (Münster: LIT Verlag, 1994), pp. 267–345. On Soviet military psychiatry more broadly, see Paul Wanke, *Russian/Soviet Military Psychiatry, 1904–1945* (London: Frank Cass, 2005).

overseeing the construction of laboratories and therapeutic training rooms. There he worked with a team of 30 researchers, including people with whom he had collaborated in Moscow and Kharkiv, whose tasks were to diagnose and treat brain injuries.[30] They worked to develop 'rational, scientifically based techniques for the rehabilitation of destroyed functions'.[31] Due to a lack of food during their first winter Luria and his family planted a garden from which they could pick mushrooms. His second wife Lana Lipchina later suffered side effects from carrying radon for use in special baths to the Urals from Moscow.[32] Luria wrote to Sergei Eisenstein describing the situation in favourable terms: 'I have a lot of staff with me and the work is going well and productively—I've even started producing some books.'[33] Luria noted that despite their modest equipment, his team's 'most important resource was dedication to the task'.[34] This was not routine scientific research but research carried out in service to the state.

Modern warfare is central to the history of the study of aphasia. Vast numbers of soldiers survived the First World War with serious brain injuries. The analysis of these casualties of modernity led to new insights into neuronal functioning and new therapeutic methods of rehabilitation. Kurt Goldstein in Germany and Henry Head in Britain dedicated themselves to healing people (overwhelmingly men) wounded in the Great War.[35] Like Luria's subsequent Soviet research, these scientists also emphasised the social function of their work, albeit in very different national contexts.

[30] Benjamin Zajicek discusses experiments conducted by Lev Perel'man and Bluma Zeigarnik investigating 'deaf-mute syndrome' at the Kisegach hospital in 'Scientific Psychiatry in Stalin's Soviet Union: The Politics of Modern Medicine and the Struggle to Define "Pavlovian" Psychiatry, 1939–1953', PhD thesis, University of Chicago, 2009, pp. 144–149. See, B.V. Zeigarnik, 'Psikhologicheskii analiz postkontuzionnykh narushenii slukha i rechi' in *Reaktivnaia postkontuzionnaia glukhonemota: Ee raspoznavanie i lechenie*, ed. L.B. Perel'man (Moscow: Medgiz, 1943), pp. 43–51. Alexander Zaporozhets and Susanna Yakovlevna Rubinstein were also among Luria's team there.

[31] Luria, *Making of a Mind*, p. 139.

[32] T.V. Akhutina, 'A.R Luriia: zhiznennyi put' [A.R. Luria: The Way of Life], *Kul'turno-istoricheskaya psikhologiia*, 2 (2012), 2–10, p. 7.

[33] Luria to Eisenstein, May 11, 1942, Moscow, Russian State Archives of Literature and Art (RGALI), Sergei Eisenstein Papers, 123-1-19321/2-13.

[34] Luria, *Making of a Mind*, p. 139.

[35] See Kurt Goldstein, *Aftereffects of Brain Injuries in War* (New York, NY: Grune and Stratton, 1942) and Henry Head, *Aphasia and Kindred Disorders of Speech* (Cambridge: Cambridge University Press, 1926).

Luria's work was explicitly indebted to their endeavours. Like Goldstein, he emphasised adaptation, readjustment and empathy, and understood wounds as both physical and psychological.[36] The rehabilitative workshops that Luria set up in the Urals (which later moved to Moscow) were a counterpart to those established by Goldstein at the Institute for Research into the Consequence of Brain Injuries in Frankfurt. Luria followed Head in observing patients over long periods of time and similarly cast his patients as 'allies' and collaborators in the therapeutic process.[37]

Luria's *Traumatic Aphasia [Travmaticheskaia Afasiia]*, published in the USSR in 1947, drew on Luria's wartime work.[38] Juergen Tesak and Chris Code give the book a privileged place in the history of aphasia: 'Of all the work on aphasia that has come from studying the victims of war, this book perhaps deserves to be recognised as the most inspired and influential.'[39] Luria discussed how the war provided unprecedented opportunities for neurologists to examine the impact of localised brain injury. His work on higher cortical functions, zones of the cortex and the restoration of brain function were built on the insights of Russian physiologist I.M. Sechenov (1829–1905).[40] The people who sustained injuries tended to be young, which improved their chances of overcoming some of the injuries' impact (a factor also emphasised by Head). The priorities were to rehabilitate people as swiftly as possible with the aim of returning them to service in the Red Army or, if their injuries were too severe, to a productive

[36] Anne Harrington discusses Golstein's research in relation to German understandings of holism. She argues that Goldstein insisted that analysis of brain-injured patients provided insights into humanity that were central to his liberal vision of society and claims that he saw 'brain-injured patients not just as an exemplary model for neurology, but also as an existential philosophy and blueprint of personal politics'. Anne Harrington, *Reenchanted Science: Holism in German Culture from Wilhelm II to Hitler* (Princeton, NJ: Princeton University Press, 1996), p. 154.

[37] On Head's wartime research, patriotic commitments and relationship to his patients as 'allies' or 'willing collaborators', see, L.S. Jacyna, *Lost Words: Narratives of Language and the Brain, 1825–1926* (Princeton, NJ: Princeton University Press, 2000), pp. 150–170.

[38] See, A.R. Luria, *Traumatic Aphasia: Its Syndromes, Psychology and Treatment*, trans. by Macdonald Critchley (The Hague: Mouton, 1970). The English text is a revised and extended version based on the Russian original: *Travmaticheskaia Afasiia: Vosstanovitelnaia terapiia* [*Traumatic Aphasia: Rehabilitative Therapy*] (Moscow: Akademiia Medititsinskikh Nauk, SSSR, 1947).

[39] Juergen Tesak and Chris Code, *Milestones in the History of Aphasia: Theories and Protagonists* (Hove: Psychology Press, 2008), p. 154.

[40] A.R. Luria, 'A.R. Luria' in *A History of Psychology in Autobiography*, vol. 6, ed. by Gardner Lindzey (Cambridge, MA: Harvard University Press, 1974), pp. 253–292, p. 268.

civilian life. The initial emphasis of treatment was focused on the healing of physical wounds. Following this stage, Luria and his team focused on rehabilitating their patients, attempting to discover ways they might recover their lost brain functions.[41] They developed therapeutic techniques, which incorporated practical tasks and encouraged patients to relearn forgotten skills.

By 1943 Luria and his team had already amassed 800 case histories. In these works by Luria the term 'trauma' [*travma*], in accordance with its etymological roots, referred to physical wounds. In *Traumatic Aphasia*, he gives little consideration to the emotional impact of war injuries or to the emotional efforts required to restore brain functions. 'Trauma' is something that happens to the tissue of the brain. Catherine Merridale refers to the 'the disappearance of individual trauma as an issue of public debate' in the Soviet Union, tying it to the state denial of the hardship, suffering and terror it unleashed on the population in the 1930s, as well as a shifting emphasis from the individual to the collective.[42] Eyes needed to be dry and fixed resolutely on the bright future.

Short case histories are littered throughout *Traumatic Aphasia*, as they are in Luria's 1948 study *Restoration of Function after Brain Injury*, but these only give minimal details about the context of the injury, the patient's personal history or their emotional state in the hospital. Instead, they all follow a similar format, providing the date and geographical location of

[41] See, for example, A.N. Leontiev and A.V. Zaporozhets, *Rehabilitation of Hand Function*, trans. by Basil Haigh (New York, NY: Pergamon, 1960).

[42] Catherine Merridale, 'The Collective Mind: Trauma and Shell-Shock in Twentieth-Century Russia', *Journal of Contemporary History*, 35, 1 (2000), 39–55, p. 48. See also, Sheila Fitzpatrick, 'Postwar Soviet Society: The "Return to Normalcy", 1945–1953' in *The Impact of World War II on the Soviet Union*, ed. by Susan J. Linz (Totowa, NJ: Rowman and Allanheld, 1985). Benjamin Zajicek challenges Merridale's account for overlooking the persistence of theories of mental trauma in Soviet psychiatry. He discusses various competing understandings of war trauma, some of which emphasised the organic (like Luria's) and some of which insisted on the continued importance of acknowledging psychological or emotional trauma. He also emphasises the different contexts in which psychiatrists were working during the war, not all of whom were as far behind enemy lines as Luria and his team, see, Zajicek, 'Scientific Psychiatry', pp. 162–167. Anna Krylova's arguments also complicates Merridale's claims by noting that although 'the central Party press ignored millions of Soviet people with lasting "spiritual" injuries, this policy was hardly the "Party line"', citing prominent literary works in which psychological trauma was a major theme (though she does concede that Soviet psychiatrists tended to be focused on the physiological more than the psychological), Anna Krylova, '"Healers of Wounded Souls": The Crisis of Private Life in Soviet Literature, 1944–1946', *The Journal of Modern History*, 73, 2 (2001), 307–331, p. 317.

injury, a description of the wound and a short account of the treatment procedure. Occasional glimpses of the affective impact of the injuries are suggested by the inclusion of short extracts from patients' first-hand accounts of their own conditions, which briefly interrupt the neutral medical tone.[43] Overall, however, as in *The Nature of Human Conflicts*, the narratives are perfunctory and de-personalised, isolating brain wounds from the subjects that bear them. The case histories are organised according to injury type, reducing patients to a proper name and military rank. The only context provided functions to explain the physical impact of the wound. These works correspond to what Luria later described as a 'classical' scientific approach, governed by 'careful, consecutive, step-by-step reasoning' but prone to reductionism.[44] In stark contrast, *The Man with a Shattered World*, a detailed case history completed decades after the war, provides a deeply personal and emotional account of the patient and his relationship to his injury, written in a literary style. The emotion the book expresses is not simply that of the patient but found in the rapport between doctor (author) and patient (protagonist), as Sacks observed in a letter to Luria: 'You could not have written … *The Man with a Shattered World* had you not loved poor Zasetsky, and felt for him the intensest fellow-feeling … had you not been able to *suffer with him*, to accompany and guide him through the *strangeness* of his extraordinary sufferings.'[45] Luria described his approach to this book as having a 'romantic' approach, in that it intended to 'preserve the wealth of living reality'.[46]

The Brezhnev era (1964–1982), during which Luria wrote his case history, saw a huge revival of the cult of the Great Patriotic War. At a time when the Soviet Union's economy and global position were struggling, past triumphs were celebrated, national pride and unity extolled.[47] Although Luria's case history is imbued with many tropes that align it with popular representations of the war produced in the period, by refusing to cast the conflict as an ecstatic victory his work strikes a more sombre note. *The Man with a Shattered World* did not attempt to deny the

[43] Luria includes an excerpt from Zasetsky's initial case history that conforms to this style in *The Man with a Shattered World*, pp. 21–22.

[44] Luria, *Making of Mind*, p. 175.

[45] Sacks to Luria, June 14, 1975, Oliver Sacks Archives, Oliver Sacks Foundation, New York City. Transliteration modified for consistency.

[46] Luria, *Making of Mind*, p. 174.

[47] See Nina Tumarkin, *The Living and The Dead: The Rise and Fall of the Cult of World War II in Russia* (New York, NY: Basic Books, 1994).

extraordinary difficulties Zasetsky faced following his injury. It thus acknowledged the lasting traumatic impact of war on individual human subjects in a manner occluded both by the dominant state-sanctioned narratives and by Luria's own more matter-of-fact medical texts from the immediate post-war period.

Despite being more circumspect on the emotional impact of war than *The Man with a Shattered World*, however, Luria's earlier texts on soldiers wounded in war were similarly addressed to the problem of overcoming or ameliorating the effects of brain injury. *Traumatic Aphasia* and *Restoration of Function after Brain Injury* were both concerned with the possibility of recovering from serious cerebral damage. Luria is clear that brain cells do not regenerate: 'the neuronal structures of the cortex, once destroyed, are incapable of regeneration.'[48] Total recovery is not a possibility; an inert scar will always remain. This did not, however, lead Luria to conclude that no improvement was possible. Unlike in Malabou's discussions of brain injury, he insisted on the possibility of healing and restoring a sense of self, however partial or incomplete the result. Rather than focusing on full recovery, Luria was concerned with compensation, substitution and reorganisation. He developed forms of treatment that slowly enabled the brain to find new ways of performing the tasks that were once performed by the damaged areas of the brain, which he described as 'Adaptation to new circumstances not to regeneration and restoration'.[49] Luria's work relied on a conception of the localisation of brain function which departed from the classic model proposed by German neurologist Carl Wernicke in its insistence on the mobility of the brain;[50] Luria understood the brain as a process rather than a product. In *Traumatic Aphasia* he described the 'dynamic localisation of functional systems in the cerebral cortex', which he likened to a constellation or mosaic of interrelating zones.[51] He contrasted the 'uniform, undifferentiated grey' appearance of the brain to the 'inconceivable complexity and differentiation' revealed through scientific

[48] Luria, *Restoration of Function after Brain Injury*, trans. by Oliver Zangwill (London: Pergamon Press, 1963), p. 32. For the Russian text, see: *Vosstanovlenie funksii mozga posle voennoi travmy* [*Restoration of Brain Function after War Trauma*] (Moscow: Izd-vo AMN SSSR, 1948).

[49] Luria, *Restoration*, p. 33.

[50] See Tesak and Code, pp. 153–158.

[51] Luria, *Traumatic Aphasia*, p. 88.

observation, insisting that the neurologist situate brains in people and people in the world.[52]

For Luria, the disruption caused by a brain injury was always understood in relation to the specific functions it precludes. As such, brain injuries were always related to a patient's former existence and to the regular flows of life from which they were wrenched:

> Lesions exclude people from everyday life and work. Restoring their functions will enable them to reintegrate into society: How may such a patient be brought back into the daily round of social activity and work? What measures must be used so that this may be done as rationally as possible?[53]

Brain injuries 'disrupt the normal life of the patient, exclude him from social intercourse and from work, and may cause irreparable damage to his intellectual life'.[54] In Luria's account, humans are distinguished from animals because they are more socially developed: 'Man's mental activity always takes place in a world of objects created during the development of society, is always directed towards them, and is frequently carried out with their aid.'[55] Greater levels of plasticity in humans also endow them a greater ability to reorient themselves to the world in the case of injury. Luria's research thus focused on restorative training using tools as external mediators between the patient and their environment. Although a long and arduous process, Luria insisted that some progress could be made: 'human activity can be restored notwithstanding the irreversible changes affecting those areas of the cortex responsible for its performance.'[56]

Restoration, adaptation and compensation in the face of permanent injury were also the focus of *The Man with a Shattered World*. Zasetsky was aware that he would never fully recover but was nonetheless committed to ameliorating his suffering. When Zasetsky first awakens in hospital he is unable to orientate himself in his environment at all: 'Right after I was wounded, I seemed to be some newborn creature.'[57] It is from the opening passages of his account that Malabou quotes, as they best demonstrate

[52] Luria, *Shattered*, p. 24.
[53] Luria, *Restoration*, p. 259.
[54] Luria, *Restoration*, p. xi.
[55] Luria, *Restoration*, p. 39.
[56] Luria, *Restoration*, p. 154.
[57] Luria, *Shattered*, p. 10.

the total discontinuity she associates with the 'new wounded'.[58] Although he may have awoken as a 'newborn creature', Zasetsky does not remain in this position of total bewilderment or what Malabou calls 'an identity without childhood'.[59] Instead, he slowly begins to find ways of anchoring himself in the world again and memories of his life before the injury begin to return: 'Afterwards, when I'd had a chance to hear words that people use in conversation or thinking, various clusters of "memory fragments" developed, and from these I began to make some sense out of the life around me.'[60]

Weiner conceives of the Second World War as one in a series of dramatic cataclysms that shook the Soviet Union. However, if the October Revolution was framed as a fundamental break with the past, a new beginning, then the Great Patriotic War, however concretely devastating, was envisioned as an attack from without that helped to consolidate an already-established national identity; the revolution smashed the existing order of things to build a new society whereas the war destroyed something that it was hoped could be rebuilt. The threat of dissolution, of rupture, was to be resisted rather than celebrated. For Zasetsky, similarly, the future was in some sense always the past; recovery is about rediscovery, about reintegration into an existing world rather than a totally new beginning. Luria explicitly described it as a 'return to life'.[61]

The process of relearning the case history describes is historically and culturally specific: the reality Zasetsky reorients himself in is a Soviet one, saturated with specific associations and ordered according to particular principles. Zasetsky is 'overjoyed' when he eventually recognises Lenin's 'familiar face' (referred to in the Russian text by the affectionate diminutive Ilicha).[62] He subsequently recalls words like sun, moon, cloud and rain, as well as his own name and the names of his family members. This implies that the image of Lenin is as taken for granted and familiar as the weather. In the period of the First Five Year Plan pedagogical experiments (discussed in earlier chapters) indicated that far from all Soviet citizens were able to recognise Lenin's face. By the time of the Second World War,

[58] Malabou, *Wounded*, p 57.

[59] Malabou, *Wounded*, p. 60.

[60] Luria, *Shattered*, p. 10.

[61] The phrase 'return to life' [*vernult'sia k zhizni*] appears in the Russian text. Luria, *Romanticheskie Esse*, p. 118.

[62] Luria, Shattered, p. 64, *Romanticheskie Esse* [*Romantic Essays*] (Moscow: Pedagogika, 1996), p. 154.

however, the newly introduced culture had become an embedded part of people's lives and consciousnesses.[63]

Zasetsky was clear that his injury made him feel useless not only in terms of his personal life, but also in relation to the life of his country. The narrative he produced, which he originally intended to give the title *I'll Fight On!*, was pervaded by militaristic vocabulary. This is even more pronounced in the Russian text; certain passages, which underline and repeat the militaristic tropes, are omitted from the English translation altogether. For instance, in the opening passages of the Russian text Luria introduces the case history by declaring: 'It is about a fight that did not lead to victory [*pobeda*], and about a victory that has not put an end to fighting.'[64] The loss Zasetsky felt was associated with patriotic feeling. He expressed concern that he was no longer 'of some service to his country' (the Russian word *narod*, more frequently translated as 'motherland', has an even more fervently patriotic emphasis).[65] The English translation also tends to omit the possessive pronoun 'our', which litters the Russian text, indicating the close identification Zasetsky feels with his nation. When presented with a copy of the daily newspaper, he is horrified to discover that he is unable to recognise 'our Pravda', later bemoaning the unfamiliarity of 'our Russian words'.[66] It is not language in general that provides his 'link with life' but a particular language, bound up with his Soviet identity and history. His sense of despair partly derives from a feeling of social disaffiliation. The specifically Soviet qualities of the text are also reflected in its form and the generic conventions employed, which mimic Zasetsky's struggle to reconnect with his former self.

[63] Serguei Alex Oushakine uses Luria's case history to illustrate a similar point, in the context of a discussion of post-Soviet nostalgia for the fallen regime, using Zasetsky's condition as a metaphor for understanding the relation to the now defunct vocabularies of state socialism. See Serguei Alex Oushakine, '"We're Nostalgic but We're Not Crazy": Retrofitting the past in Russia', *Russian Review*, 66, 3 (2007), 451–482. He also employs the metaphor of aphasia in 'In the State of Post-Soviet Aphasia: Symbolic Development in Contemporary Russia', *Europe-Asia Studies*, 52, 6 (2000), 991–1016.

[64] Luria, *Romanticheskie Esse*, p. 105. In Russian, the word '*pobeda*' is also a short-hand term to refer to Soviet victory in the Second World War specifically. A similar sentence appears at the end of the Russian text (immediately preceding the epilogue) that is also omitted from the English (Compare, *Shattered*, p. 159 and *Romanticheskie Esse*, p. 232). Compare also, *Romanticheskie Esse*, p. 118 with *Shattered*, p. 17 which omits a short paragraph describing Zasetsky's 'difficult struggle'.

[65] Luria, *Shattered*, p. 35 and Luria *Romanticheskie Esse*, p. 131.

[66] See, Luria, *Romanticheskie Esse*, p. 154. In the English text 'our newspaper Pravda' is translated as 'Pravda' and 'our Russian words' as 'words'. Luria, *Shattered*, p. 64.

A World Shattered and Remade: Between Spontaneity and Consciousness

Sacks conceived of case histories, or what he called 'Clinical Tales', as narratives that strive to reorient the self in a world experienced as broken; they are concerned with 'organisation and chaos, order and disorder'.[67] Zasetsky framed his struggle in similar terms, as an attempt to make 'the bits and pieces to add up to a coherent whole'.[68] This is also captured by the Russian title of the case history: *A World Lost and Re-gained* [*Poteryannyi i vozvrashchennyi mir*]. For Sacks, a clinical tale aims to describe the relation of the patient's altered world to 'our world'.[69] But whose world is 'ours'? Notions of chaos and disorder rely on a conception of organisation and order, but such apparently stable entities are historically contingent. As we have seen, Zasetsky's world was ordered in a particular way that was partially determined by his Soviet context. The worlds Sacks and Luria attempted to reintegrate their patients back into were ordered differently, and so were the narrative styles they employed to convey their patients' conditions.

Luria was an avid reader of detective novels. Oliver Sacks and Jerome Bruner recalled that they were both implored to supply him with works by American crime writers including Ngaio Marsh and Nicolas Freeling. Sacks recalled that Luria's favourite Sherlock Holmes book was the *Seven Percent Solution*, in which a cocaine-addicted Holmes is in analysis with Freud.[70] Luria explicitly likened the role of the psychologist to that of a 'criminal investigator'.[71] Indeed, his experiments with criminals included

[67] Oliver Sacks, 'Clinical Tales', *Literature and Medicine*, 5 (1986), 16–23, p. 18. For a persuasive critique of this kind of argument, see, Angela Woods, 'The Limits of Narrative: Provocations for the Medical Humanities', *Medical Humanities*, 37, 2 (2011), 73–78, p. 74.

[68] Luria, *Shattered*, p. xxi.

[69] Sacks, 'Clinical Tales', p. 18.

[70] Michael Cole, Jerome Bruner and Oliver Sacks, 'A Dialogue about Alexander Luria', *Dubna Psyanima Journal*, 1 (2013), 41–49. In 1975 Luria requested Sacks send him paperbacks by Nicolas Freeling 'who to me is one of the real best detective, or psychological detective, writers', requesting *Double Barrel, Criminal Conversation, The Ring of the Rainy Country, The Dresden Green* and *Strike Out Where Not Applicable*. Luria to Sacks, August 4, 1975, Oliver Sacks Archives. Sacks in turn described Luria as a 'great Detective who is, equally and necessarily, a great Reconstructor'. Sacks to Luria, December 24, 1975, Oliver Sacks Archives, Oliver Sacks Foundation, New York City.

[71] K.E. Levitin, 'The Soul's Frail Dwelling House', *Journal of Russian and East European Psychology*, 5, 36 (1998), 23–45, p. 27.

in his 1932 book *The Nature of Human Conflicts* (discussed in the opening chapter) even saw him working directly with the police. He wrote to Sacks in 1975:

> The method of recovering the whole structure of events, putting the pieces together (in no way making the opposite—splitting the whole to pieces and reducing the most complicated whole to its parts). That is what we have in both—in medicine and in detective stories.[72]

In an essay analysing popular detective fiction, 'The Hotel Lobby' (1925), Siegfried Kracauer claims that both the detective and the detective novel seek to 'construct a whole out of the blindly scattered elements of a disintegrated world'.[73] Luria's case history set out to perform a similar task.

Kracauer also analyses the detective novel as a product of mass culture that provides clues to the mysteries of the particular society that produced it. As he observes in 'The Mass Ornament' (1927):

> The position that an epoch occupies in the historical process can be determined more strikingly from an analysis of its inconspicuous surface-level expressions than from that epoch's judgments about itself.[74]

Kracauer declares that the detective novel 'discloses the secret of a society bereft of reality'.[75] He sees the alienation of capitalist society reflected in this genre of popular fiction. Kracauer discusses the hotel lobby as the location that defines detective fiction—a space illuminated by harsh artificial light, and inhabited by spectral, disconnected figures whose apathy and inertia might be linked to Malabou's characterisation of the 'new wounded', which she also explicitly characterises as a distinctly capitalist form of subjectivity.[76] Luria's international contacts and ability to read in English gave him access to popular Western fiction,[77] but the dialectical

[72] Luria to Sacks, July 1, 1975, Oliver Sacks Archives, Oliver Sacks Foundation, New York City.

[73] Siegfried Kracauer, 'The Hotel Lobby' in *The Mass Ornament: Weimar Essays*, trans. by Thomas Y. Levin (Cambridge, MA: Harvard University Press, 1995), pp. 173–188, p. 175.

[74] Siegfried Kracauer, 'The Mass Ornament' in *The Mass Ornament*, pp. 75–86, p. 75.

[75] Siegfried Kracauer, 'Hotel Lobby', p. 175.

[76] See, particularly, Malabou, *What Should We Do with Our Brain?*

[77] On the fate of detective fiction in the Soviet Union, see Natalia Ilyina and Arkadi Adamov, 'Detective Novels: A Game of Life', *Soviet Literature*, 3 (1975), 142–150 and Robert Russell, 'Red Pinkertonism: An Aspect of Soviet Literature of the 1920s', *SEER*, 60 (1982), 390–412.

tension between woundedness and healing that animated *The Man with a Shattered World* bore much closer resemblance to a popular Soviet literary genre remote from American detective fiction: the socialist realist novel. Kracauer's approach to the products of mass culture might be usefully transposed into the Soviet context: attending to the 'surface-level' of Luria's text provides insights into the epoch and society of which it was a product.

Socialist realism was installed as the official literature of the Soviet Union following the Soviet Writers' Congress in 1934, the year of Vygotsky's death. The term itself had been in circulation for a couple of years already, and the condemnation of 'Freudism', Formalism and modernism were already taken for granted when the conference began. Katerina Clark discusses the socialist realist novel as a form of ritual, understood as a cultural form and social act through which cultural meaning and national myths were generated, consolidated and sustained. Socialist realist novels follow a distinct master plot, whose formula, she argues, remained remarkably consistent from the 1930s through to the Brezhnev era, with even illegally circulated *samizdat* literature tending to conform to the officially established structure.[78] The Soviet novel's master plot was 'the literary expression of the master categories that organise the entire culture'.[79] As Jochen Hellbeck discusses in his analysis of Soviet diarists, *Revolution on My Mind*, a broadly Leninist understanding of society's progression towards communism and the individual's place in that struggle also structured many Soviet citizen's private reflections: 'To truly live meant to rise above selfish pursuits and devote one's life to

[78] On literary debates during and in the immediate aftermath of the Second World War and the adaptation of socialist realism to wartime concerns, see Anna Krylova, '"Healers of Wounded Souls": The Crisis of Private Life in Soviet Literature, 1944–1946', *The Journal of Modern History*, 73, 2 (2001), 307–331. She argues that 'the privileging of narratives of overcoming' both in literature and in the press consigned 'by default other psychological invalids unable to overcome their traumas to the realm of the unrepresentable and, thus, to symbolic death', p. 316. It is my contention that Zasetsky combined a failure to recover or overcome with a determination to do so which both rendered his condition legible and pointed to the limits of the kinds of narratives of overcoming considered by Krylova.

[79] Katerina Clark, *The Soviet Novel: History as Ritual* (Chicago, IL: University of Chicago Press, 1981), p. 14. Irina Gutkin similarly discusses the 'homogeneity of cliched language' employed within Soviet literature, a system which created a 'system of linguistic bondage' in which '[e]very word is firmly bonded, spatially and temporally, to ideological tenets'. Irina Gutkin, *The Cultural Origins of the Socialist Realist Aesthetic*, 1890–1934 (Chicago, IL: Northwestern University Press, 1999), pp. 68–69.

society and history, to remake accursed, backward Russia through the power of personal example and an unflinching orientation toward the future.'[80] Zasetsky's accounts of his illness are consistent with Hellbeck's characterisation of Soviet diary-writers: writing was framed as an arduous pursuit which functioned as a space for self-interrogation, where the lonely and inadequate individual strove to overcome their personal deficiencies in order to merge with the collective.

According to Clark, the hero of the socialist realist novel embodies the Marxist-Leninist account of history. The questing protagonist undergoes a transformation, which she characterises as a working out of the dialectic between spontaneity [*stikhinost'*] and consciousness [*soznatel'nost'*], the movement famously described by Lenin in *What is to be Done?* (1902). Despite being intemperate and chaotic, for Lenin, spontaneity is a prerequisite for the progression towards communism. The movement from spontaneity to consciousness is a progression from chaotic resistance to organised revolutionary struggle. In Clark's account, this movement had a counterpart in the socialist realist hero:

> The great historical drama of struggle between the forces of spontaneity and the forces of consciousness is unfolded in a tale of the way one individual mastered his wilful self, became disciplined, and attained to an extrapersonal identity … the hero achieved greater harmony within himself and in relation to his society.[81]

The Man with a Shattered World overwhelmingly corresponds to the master plot of the socialist realist novel.[82] The case history describes a heroic

[80] Jochen Hellbeck, *Revolution on My Mind: Writing a Diary Under Stalin* (Cambridge, MA: Harvard University Press, 2006), pp. 362–363. Sean Guillory's analysis of first-hand accounts of the Civil War suggests that the ideological narratives of the state were less prevalent in this early period. Instead, he argues that these memoirs reveal: 'confusion, personal loss, hardship, physical suffering, and fear in the face of death'. Sean Guillory, 'The Shattered Self of Komsomol Civil War Memoirs', *Slavic Review*, 71, 3 (2012), 546–565, p. 547.

[81] Clark, pp. 16–17.

[82] Rebecca Reich similarly claims that socialist realist tropes structured reports and publications written by psychiatrists involved in the pathologisation of dissidents during the late Soviet period, arguing that such works conformed to an 'established aesthetic framework', Rebecca Reich, *State of Madness: Psychiatry, Literature and Dissent after Stalin* (DeKalb, IL: Northern Illinois University Press, 2018), p. 27.

man—Luria explicitly describes him as such throughout[83]—struggling with almost superhuman effort and determinacy to overcome his own infirmity. Zasetsky is committed to overcoming chaos and re-establishing the order of his broken life in order 'to impose some order and sense of continuity'.[84] 'This book', Luria asserts in a passage omitted from the English translation, 'is about a heroic struggle for life'.[85] Nature often appears as the antagonist in socialist realist narratives—dam building, farm collectivising and factory constructing in hostile environments all feature prominently in novels of the genre. For Zasetsky, however, the nature being overcome is not external but internal; Zasetsky is at war with his own brain.

Lilya Kaganovsky identifies two conflicting yet coexisting images of masculinity operating within Stalinist culture: the healthy, virile, strong figure familiar from socialism realist statues and paintings, and the damaged, scarred and mutilated male body that appeared in cinematic and literary works of the period. She sees superimposed over the image of 'extravagant virility' presented in the visual arts, the heroic invalid—'limping, bandaged, bedridden, grounded, unwilling or unable to marry'—a staple archetype in Soviet literature and film which punctures the idealised image of masculinity[86]: 'The heroic invalid *is* the exemplary Stalinist subject.'[87] Zasetsky conforms to this archetype. Kaganovsky reads the coexistence of these conflicting images of masculinity in psychoanalytic terms, as a fantasy of 'radical dismemberment', representing the fundamental gap between real and ideal gender norms.[88] She frames this in Lacanian vocabulary: her heroes live with lack, staging the incom-

[83] Although the English text does describe Zasetsky as a hero, this aspect of Luria's narrative style is far more pronounced in the Russian text. Compare, for example, Luria, *Romanticheskie Esse* (pp. 105–106) in which the words 'hero' or 'heroic' appear seven times but only once in the English (Luria, *Shattered*, p. 15). At a later point 'our hero' (Luria, *Romanticheskie Esse*, p. 231) is translated simply as 'this man' (Luria, *Shattered*, p. 157).

[84] Luria, *Shattered*, p. xix.

[85] Luria, *Romanticheskie Esse*, p. 105.

[86] Lilya Kaganovsky, *How the Soviet Man was Unmade: Cultural Fantasy and Male Subjectivity under Stalin* (Pittsburgh, PA: University of Pittsburgh Press, 2008), p. 3. On masculinity and the Soviet experience of the Second World War, see Catherine Merridale, *Ivan's War: Life and Death in the Red Army, 1939–1945* (New York, NY: Metropolitan Books, 2006).

[87] Kaganovsky, p. 129. Interestingly, Jacyna notes that the archetypal aphasic person is also male, *Lost Words*, p. 7.

[88] Kaganovsky, p. 3.

mensurability of penis and phallus. But the gap between reality and ideal is not simply a question of gender as Kaganovsky implies; it is precisely what animated narratives of the experience of building socialism. Unlike nineteenth-century realism, socialist realism was not strictly mimetic; it was not concerned with simply depicting the present as it really was. Socialist realism was never intended as a mirror of the present but more like a window into the future, juxtaposing an idealised image of current existence with a prefiguration of the bright Communist world to come. Régine Robin characterises socialist realism as an 'impossible aesthetic' whose impossibility derives from the fundamental incompatibility she discerns between the conflicting temporalities at work within socialist realist narratives, which she describes as 'the impossible inscription of the description-prescription oxymoron'.[89] From this tension between the what-is and the what-ought-to-be, the narrative derives its relentless forward-moving drive; the novel is always dynamic, with an emphasis on the ongoing struggle to build communism, to bring a new reality into being. The fantasy (future) of unity and reconciliation is combined with the reality (present) of dismemberment. Radical dismemberments paradoxically propel the hero forwards; he (and it is almost always a 'he')[90] strives to overcome the obstacles in the external world as well as those presented by his own body.

The most striking parallel with *The Man with a Shattered World* is Nikolai Ostrovsky's prototypical *How the Steel was Tempered*, first published in serial form between 1932 and 1934, just prior to the 1934 Soviet Writer's Congress. The novel tells the story of revolutionary hero Pavel Korchagin who joins the Red Army during the Civil War, tracing the young protagonist's progress from passive individual to collective subject. Pavel is severely injured in battle but his disability does nothing to quell his revolutionary fervour. At the end of the novel, though bedridden and blind, Pavel remains 'determined to resume his place in the ranks of the

[89] Régine Robin, *Socialist Realism: An Impossible Aesthetic*, trans. by Catherine Porter (Palo Alto, CA: Stanford University Press, 1992), p. 297. Clark describes this phenomenon as 'modal schizophrenia', p. 38. Gutkin glosses it in the following terms: 'The socialist realist chronotope projected into mass consciousness the dichotomous myth of the beautiful ideal future and the already beautiful present', p. 80.

[90] For a late Soviet attempt to both address and redress the exclusion of women from dominant wartime discourses, see Svetlana Alexievich, *The Unwomanly Face of War*, trans. by Richard Pevear and Larissa Volokhonsky (London: Penguin Books, 2017) (originally published in Russian in 1985).

builders of the new life'.[91] Like Zasetsky, he does this by writing an account of his struggles. The novel concludes when he receives a telegram from Leningrad confirming that his work will be published, returning him 'to the fighting ranks and to life'.[92] Zasetsky's writing played a similar role and he expressed himself in the embattled vocabulary of the socialist realist hero: 'I'm fighting to recover a life I lost when I was wounded.'[93] In both cases the text, the process of writing and the writing subject cannot be disentangled. Ostrovsky's own life famously paralleled that of his mutilated hero. The novel's torturous composition formed an integral part of its symbolic status, elevating writing to a heroic and revolutionary act.[94] Zasetsky attempted to achieve consciousness through writing, and the case history bears witness to the strenuous therapeutic process of which it was a product.

The Writing Cure

Psychoanalysis is a talking cure; the treatment proceeds through a peculiar form of conversation in the consulting room, configured by temporal and monetary constraints. Friedrich Kittler argues that the psychoanalytic encounter should be understood as a 'pure realm of hearing'.[95] According to Kittler, psychoanalysis has 'no vague parallels to film; it has much more precisely learned the lesson of technological sound recorders',[96] which originated 'not coincidentally' at the same time.[97] The acoustical data of the consulting room is not recorded by a machine but listened to by an

[91] Nikolai Ostrovsky, *How the Steel was Tempered*, vol. 2, trans. R. Prokofieva (Moscow: Foreign Languages Publishing House, 1959), p. 375. Parallels between Luria's work and Wanda Wasilewska's 1945 novel *Simply Love* with its war-wounded protagonist 'torn between his insistence on his nonexistence and his desire to reconstruct the link between his past identity and that of the present or future' could also be drawn. See Krylova, p. 323.

[92] Ostrovsky, p. 382.

[93] Luria, *Shattered*, p. 16.

[94] See, Semen Adol'fovich Tregub, *The Heroic Life of Nikolai Ostrovsky* (Moscow: Foreign Languages Publishing House, 1957).

[95] Friedrich Kittler, *Discourse Networks, 1800/1900*, trans. by Michael Metteer with Chris Cullens (Stanford, CA: Stanford University Press, 1990), p. 284. Sander Gilman makes a similar observation. See Sander Gilman, *Seeing the Insane* (New York, NY: John Wiley, 1982), p. 223.

[96] Kittler, *Discourse Networks*, p. 284.

[97] Kittler, *Gramophone, Film, Typewriter*, trans. by Geoffrey Winthrop-Young and Michael Wutz (Stanford, CA: Stanford University Press, 1999), p. 37.

imperfect human analyst who transforms the material of the session into written words from memory: 'This is the relation between analytic practice and theory. What would have disturbed free-floating attention during the session later occurs after all: Freud reaches for his pen'.[98] Freud did not take notes during analytic sessions, but relied on his memory alone, as he notes in 'Dora' (1905): 'the record is not absolutely—phonographically—exact.'[99] Whereas the phonograph records noises 'regardless of their so-called meaning',[100] the writing analyst, even if they select to write nonsense, always selects and, concomitantly, they always omit. Kittler's critique of the psychoanalytic method is grounded in a concern with the transformations and distortions that occur when one medium (speech) is transposed into another (writing).

Kittler's discussion is primarily concerned with how the written case history is produced, but a body of feminist secondary literature on the Freudian case history has focused on the issue of who produced the written text, and attempted to redress the imbalance balance between writing analyst and speaking analysand by reinserting the analysand's voice into the history of the case.[101] In a similar vein, Malabou accuses Luria of muffling his patient's experience by smothering it with his own written account, claiming that Luria falsely imposes clarity on his fragmented patient: 'narrative work is a clinical gesture'.[102] Unlike Freud's case histories as discussed by Kittler, which transpose one medium into another, however, Luria composed his case history from texts written by Zasetsky, the composition of which was central to the therapeutic process. Sacks noted admiringly of Luria's therapeutic approach in a letter that it seemed to him that Luria did 'not do things to patients, but explore[d] with them (and for them) the exact qualitative structure of their disordered judgements/actions, and ways in which

[98] Kittler, *Discourse Networks*, p. 285.

[99] Sigmund Freud, 'Fragment of an Analysis of a Case of Hysteria' ['Dora'], *Standard Edition*, vol. 7 (London: Hogarth Press, 1953), pp. 3–124, p. 10.

[100] Kittler, *Gramophone, Film, Typewriter*, p. 85.

[101] See, for example: Lisa Appignanesi and John Forrester, *Freud's Women* (New York, NY: Basic Books, 1992), *In Dora's Case: Freud-Hysteria-Feminism*, ed. by Charles Bernheimer and Claire Kahne (New York, NY: Columbia University Press, 1985) and Hannah S. Decker, *Freud, Dora, and Vienna 1900* (New York, NY: The Free Press, 1991). See also, Hélène Cixous, 'Portrait of Dora', trans. by Anita Barrows, *Gambit International Theatre Review*, 8, 30 (1977), pp. 27–67, p. 54.

[102] Malabou, *Wounded*, p. 54. L.S. Jacyna similarly discusses the status of the aphasic subject in scientific case histories, concluding that historically such subjects had a 'lack of power to influence the acts of representation within which they were entwined'. See, Jacyna, p. 6.

these may be reconstituted in their original form, or, if this is impossible, in some surrogate form'.[103] Following Kittler's argument, by operating within the same medium as his patient, Luria's method could be said to present his patient in a less mediated (though certainly not unmediated) fashion than the psychoanalytic case history. Although the patient's experience is still framed by the neurologist, his words are represented in the case history in a manner the psychoanalytic method precludes.

Kittler extends his discussion of the limitations of the human analyst by pointing to the failure of Freud's case histories to capture the 'white noise' of free association, which, he claims, imposes order on chaos. Free association is characterised by a lack of calculation, of organisation, of coherence; the analysand is encouraged to relate whatever is on the 'surface' of their mind however meaningless, trivial or absurd-seeming.[104] In producing a case history, the analyst alone is responsible for reproducing the unruly utterances (and silences) of the analysand, but Luria's techniques and goals were very different from those of free association as he emphasised the voluntary over the involuntary. Luria encouraged Zasetsky (who, unlike the subjects of Freud's case histories, was not given a pseudonym) to use writing to reorder his own broken life. In the opening pages of the case history he explicitly declared that Zasetsky should be considered the book's author: 'It is not false modesty on my part to wish no credit for this book. The real author is its hero.'[105] It is not the doctor but the patient who took responsibility for piecing things back together, inserting shards of memory into a coherent narrative: 'His only material consisted of fragmentary recollections that came to mind at random. On these he had to impose some order and sense of continuity.'[106]

Luria frequently employed writing as a form of therapy with brain-injured patients. Initially, he claimed that his patients experienced their memories as 'pieces of a jigsaw puzzle piled up haphazardly' before them.[107] Luria devised

[103] Sacks to Luria, February 22, 1974, Oliver Sacks Archives, Oliver Sacks Foundation, New York City.

[104] For a definition of free association and an account of its pre-psychoanalytic origins, see J. Laplanche and J.B. Pontalis, *The Language of Psychoanalysis*, trans. by Daniel Nicholson-Smith (London: Karnac Books, 1988), pp. 169–170.

[105] Luria, *Shattered*, p. xix.

[106] Luria, *Shattered*, p. xx.

[107] Luria, *Making of Mind*, p. 153. The jigsaw metaphor is repeated in a letter from Sacks to Luria who notes that Zasetsky's 'shattered cerebral jigsaw' could be reassembled. Sacks to Luria, June 30, 1975, Oliver Sacks Archives, Oliver Sacks Foundation, New York City.

a writing exercise to assist these people in completing the puzzle. The patient would write down 'the fragments of narration they remember on separate pieces of paper', lay the pieces in front of them and rearrange them into a 'coherent order'.[108] In one experiment aimed at restoring a sense of narrative continuity the patient was given a Tolstoy story to read and asked to reproduce aspects of it, in another the patient was presented with shuffled index cards relating to their own injury entitled 'The Story of How I Was Wounded' and asked to place the cards in chronological order.[109] Using similar methods, writing provided Zasetsky with his only 'link with life, his only hope of not succumbing to illness but recovering at least part of what had been lost'.[110] Remastering his lost language allowed Zasetsky to reconnect to humanity. Writing acted as a mediator between patient and world; the process of composing narratives was used to weave the frayed strands of his life back into the fabric of society. Writing was crucial to the therapeutic process, rather than an act performed by the psychologist *post factum*. Furthermore, in contrast to the analysand who reclines on the couch as a means of encouraging the relaxation necessary for free association, Luria described Zasetsky's process of writing as a laborious and deliberate pursuit. Initially, Zasetsky could not even remember how to use a pencil or form letters on a page. Although he gradually remastered the ability to write, producing words remained torturously difficult: 'every word he recalled, every thought he expressed, required the most excruciating effort. When his writing went well, he managed to write a page a day, two at most, and felt completely drained by this.'[111] Yet it was this constant battle with linguistic expression that gave Zasetsky's life meaning: 'This work caused him no end of despair, but the hope that had impelled him to write carried him along.'[112]

[108] Luria, *Making of Mind*, p. 154.

[109] A.R. Luria, *Human Brain and Psychological Processes*, trans. by Basil Haigh (New York, NY: Harper & Row, 1966), pp. 266–274. This book also contains many short case histories of people injured in war similar in length and form to those included in *Traumatic Aphasia* and *Restoration of Function after Brain Injury*.

[110] Luria, *Shattered*, p. xx.

[111] Luria, *Shattered*, p. xx.

[112] Luria, *Shattered*, p. 76.

The frenzied approach Zasetsky took to writing and the fervent application that the case history returns to throughout, did not, however, succeed in producing a neatly ordered document:

> I tried to remember whatever I could with that battered memory of mine and write it as a true story, just as a writer would. But when I started I realised I'd never be able to do that since I didn't have enough of a vocabulary or mind left to write well.[113]

Luria based *The Man with a Shattered World* on 3000 pages of written notes composed by Zasetsky over a period of over 25 years. He evocatively described the material output yielded by the therapeutic process: 'Before me is a pile of notebooks: some of them faded, makeshift affairs dating from the war years; others, thick, oilcloth-covered books covering the recent past.'[114] Although the transformation that occurred in turning this raw material into a case history may not be strictly analogous to the process of composing a psychoanalytic case history, which relies on the analyst's memories of the spoken analytic encounter, it was still Luria who edited, arranged and commented upon the huge unwieldy mass of written material Zasetsky produced. Unlike in a psychoanalytic consulting room the therapist encouraged his patient to order his thoughts, but this nonetheless proved impossible. Indeed, Zasetsky was only capable of writing automatically, 'without thinking', which suggests that his method of producing words was not as remote from free association as his descriptions of his intense application to the task initially seem to imply.[115] Zasetsky might have been driven to heal, to restore himself to the world, but the vast accumulation of notes that he produced suggest that his efforts were always undercut by the impossibility of doing so: 'I'd try to dig up another idea and find suitable words for it, and I'd write these down on various scraps of paper ... what a torture it was.'[116] He forgot the most simple words, produced 'disjointed phrases' and found rereading his own words extremely difficult.[117] In the pursuit of making sense of his life, Zasetsky piled words upon words, pages upon pages and notebooks upon notebooks; the pursuit of consciousness was undercut by the spontaneity of experience.

[113] Luria, *Shattered*, p. 79.
[114] Luria, *Shattered*, p. xix.
[115] Luria, *Shattered*, p. 77.
[116] Luria, *Shattered*, pp. 79–80.
[117] Luria, *Shattered*, p. 78.

Ultimately, for all its strenuous exertions to pull the disparate pieces together, the work and the patient it describes remained in pieces:

> I wanted terribly to write this story but I worked so hard at it, I finally felt sick—both from my head wound and the endless job of writing about it. It's been an enormous strain (still is). I work at it like someone with an obsession.[118]

In 'Dora' Freud claims that an analysand's initial account of their experiences 'may be compared to an unnavigable river whose stream is at one moment choked by masses of rock and at another divided and lost among shallows and sandbanks ... leaving gaps unfilled, riddles unanswered'.[119] It is only by the end of the treatment that 'an intelligible, consistent and unbroken case history' will emerge.[120] Freud, however, is quick to undermine this claim. The case history is, after all, entitled 'Fragment of an Analysis of a Case of Hysteria' and his patient cut off her analysis. Despite his efforts, Luria, like Freud, was incapable of replacing the unnavigable river with a straight canal (a conclusion no socialist realist novel would tolerate). Luria's case history should not be condemned, as per Malabou's argument, for failing to communicate the peculiar qualities of his patient's condition, however, because the text does not completely banish the unruly quality of his patient's writings. The text might not be a transparent representation of Zasetsky's condition but its fragmented form bears witness to the relationship between the patient and his doctor, and between Zasetsky's present experiences and imagined healthy future. The gesture of ordering the material into a case history mimics the structure of the treatment, and ultimately the fragmentary, non-linear structure of the former, does justice to the unfinished, interminable quality of the latter.

Returning once more to the form of the text, it is possible to discern moments in which the progressive thrust of the socialist realist narrative is undermined by Zasetsky's constant digressions and regressions. Luria's contemporary, the Soviet literary scholar Mikhail Bakhtin, argued that language is always internally split.[121] He acknowledged that language is

[118] Luria, *Shattered*, p. 81.

[119] Sigmund Freud, 'Dora', p. 16.

[120] Freud, 'Dora', p. 18.

[121] On overlaps between Bakhtin's ideas and Vygotsky's and personal connections within the Soviet intelligentsia, see Caryl Emerson, 'The Outer Word and Inner Speech: Bakhtin, Vygotsky, and the Internalization of Language', *Critical Inquiry*, 10, 2 (1983), 245–264 and Craig Brandist, *The Bakhtin Circle: Philosophy, Culture and Politics* (London: Pluto Press, 2002).

'ideologically saturated',[122] claiming that the apparently rigid structures of what he called 'unitary language' always operate in the midst of heteroglossia; a tendency to unite is constantly undermined by a tendency to disassemble. The unitary language and heteroglossia are in constant interaction—the former imposing limits, tending towards stasis and ossification, the latter always fighting to resist these constraints. The unitary language is centripetal, it keeps the language moving uniformly along a prescribed path, whereas heteroglossia is centrifugal, it has a decentralising tendency. For Bakhtin, every utterance is 'a contradiction-ridden, tension-filled unity of two embattled tendencies in the life of language'.[123] The contradictory qualities of language described by Bakhtin recall the movement traced in Freud's *Beyond the Pleasure Principle* (an essay for which Luria had co-written an introduction in 1925), in which Eros or the life instincts draw things together while the death drive pulls them apart. These contradictions and tensions also animate *The Man with a Shattered World*.

Zasetsky's injury produced linguistic and perceptual disturbances that the treatment attempted to overcome without ever fully succeeding. Zasetsky sought to reintegrate into the existing world from which his injury excluded him, a process the case history formally and linguistically mimicked. Ultimately, however, Zasetsky was operating within a particular ideologically saturated discourse that the condition being described refused to be fully contained by. The specificities of the cerebral condition of the patient cracked open the official unitary language with its emphasis on wholeness and resolution. Heteroglossia intruded, unravelling the centripetal qualities of the text's socialist realist unitary language.

The Man with a Shattered Brain: The History of a World Wound

Malabou challenges the assumption that narrative can function to reunite wounded people with their former selves. Instead she asks: 'What mirror could reflect a brain?'[124] Although she insists that the 'new wounded' cannot be healed, she does suggest that their experience of total detachment could be represented. Luria's case history, Malabou argues, fails to

[122] Mikhail Bakhtin, 'Discourse in the Novel', *The Dialogic Imagination: Four Essays*, ed. by Michael Holquist, trans. by Michael Holquist and Caryl Emerson (Austin, TX: University of Texas Press, 1981), p. 271.

[123] Bakhtin, p. 272.

[124] Malabou, *Wounded*, p. 55.

do justice to the 'rupture of narrativity' that she assumes characterises the psychic condition of the patient being described.[125] Instead, she identifies Samuel Beckett's 'theatre of absence' as the literary form most capable of representing the 'affective impoverishment and destructive metamorphosis' that she insists is characteristic of the new wounded.[126] Beckett's writings, she claims, are uniquely capable of capturing the 'blank spaces that emerge when the network of connections is shredded'.[127] Malabou proposes that Beckett's work might be suited to conveying the subjectivity of the new wounded because of its 'interruption, pauses, caesuras'.[128] In practice, however, her analysis focuses on content rather than form; she is more concerned with the kind of subjects she claims Beckett describes, than in his use of language *per se*. For example, she draws heavily on the recent work of contemporary popular neuroscientist Antonio Damasio, favourably remarking upon his predilection for Beckett, without pausing to consider the literary form Damasio employs, in which the matter-of-fact voice of the doctor takes precedence over those of his patients in a far more pronounced and conventional way than in Luria's case history, where the first-person perspective switches between doctor and patient. The subjects Damasio describes may conform to Malabou's singular definition of the 'new wounded' but she does not ask why the narrative style of his case histories, which are written in continuous prose without interruptions, pauses or caesuras, bears no trace of his professed literary influences.[129]

Unlike Damasio's cases, *The Man with a Shattered World* proceeds in a jerky, jolted fashion. Although concerned with establishing unity and coherence, Luria does not present the wounded subject as already healed nor is the text unified and coherent. Luria employs two interweaving narrative voices, switches between various tenses and genres and jumps backwards and forwards in time in order to mimic the patient's non-linear experience. By relying on large chunks of first-person testimony, it is written from the perspective of terrified disorientation, as felt by Zasetsky:

[125] Malabou, *Wounded*, p. 55.
[126] Malabou, *Wounded*, p. 55.
[127] Malabou, *Wounded*, p. 55.
[128] Malabou, *Wounded*, p. 55.
[129] For examples of Damasio's case histories, see Antonio Damasio, *The Feeling of what Happens: Body, Emotion and the Making of Consciousness* (London: Harcourt, 1999).

> When I came out of the bathroom, I forgot which way I had to turn to get back to my room. So I just started walking, dragging myself along. Suddenly I banged my side against the door. [...] I was amazed that this had happened. [...] I turned in the other direction and fell, because I got confused again and didn't know which way to walk. Suddenly the words right, left, back, forward, up, and down occurred to me, but they weren't any help since I didn't really understand what they meant.[130]

The book is full of surreal and bewildering passages that might indeed come from the pages of Beckett: spoons appear as 'peculiar bits of space',[131] familiar towns become alien labyrinths, heads might be as large as tables one minute and as small as chicken heads the next, hands and legs vanish and must be 'hunted' for, and written language appears as a mysterious series of squiggles. Zasetsky sees everything as though through a swarm of tiny flies. Luria writes that the objects Zasetsky 'saw no longer appeared stable. They would glimmer fitfully and become displaced, making everything appear as if it were in a state of flux.'[132] Words appear as though they had been 'gnawed, plucked around the edges, and what's left are scattered points, quills, or threads that flicker like a swarm'.[133] Objects confront him as completely mysterious, 'alien, colourless and fragmented'.[134] For Zasetsky, vision takes on a monstrous life of its own; it not only obscures reality but is actively nibbling away at it, wearing it down: 'I see only a grey mist in which spots, threads, and little bodies seem to shift and flicker back and forth.'[135]

Immediately after regaining consciousness Luria describes asking his patient to recount the moment of his injury: '"Well, you see ... it's, it's ... a long time already ... must be two, three. ... What's the word? ..."'[136] This last question, as Laura Salisbury discusses, recalls Beckett's final work 'What Is The Word', written after its author suffered a stroke that resulted in temporary aphasia. Salisbury describes a work 'scattered with stuttered dashes, abrupt elisions, compulsive repetitions and controlled echoes that inhabit an uncanny hinterland between the voluntary and

[130] Luria, *Shattered*, p. 53.
[131] Luria, *Shattered*, p. 37.
[132] Luria, *Shattered*, p. 38.
[133] Luria, *Shattered*, p. 38.
[134] Luria, *Shattered*, p. 99.
[135] Luria, *Shattered*, p. 75.
[136] Luria, *Shattered*, p. 17.

the involuntary'.[137] Unlike Beckett, Luria attempted to overcome the disruptions. Writing had a therapeutic function, providing Zasetsky with a means to reorient himself in a world experienced as broken. Nonetheless, despite the efforts of its author(s), *The Man with a Shattered World* also evokes an uncanny hinterland between voluntary and involuntary; the dialectic between woundedness and healing is never fully resolved.

The text is riven with breaks that correspond to gaps in Zasetsky's memory. Contrary to Malabou's criticisms, Luria's translator claims that 'Luria has scrupulously preserved the repetitions and inconsistencies which are symptoms of the patient's condition'.[138] Elisions punctuate the description of his last memories before being wounded: 'Under fire, I jumped up from the ice, pushed on … toward the west … there … and ….'[139] Luria also reproduces the pauses and non-sequiturs that dotted Zasetsky's speech. In its form this recalls Gilles Deleuze's discussion of Beckett's writing, which he described by way of a neurological metaphor (borrowed from Beckett's 1937 letter to Axel Kaun)[140]: 'Beckett clamours for a style that would proceed through a perforation and proliferation of tissue.'[141] Malabou briefly mentions Deleuze's characterisation of Beckett as a writer of exhaustion without noting that in Deleuze's discussion the exercise of exhausting the possible in Beckett's writing (achieved through exhaustive combinations of words) does not result in the kind of barren qualities she associates with the 'new wounded' at all. Instead, Deleuze argues, foregrounding the strangeness of language allows for a kind of desubjectivised affect to seep back into Beckett's texts. Drilling through language results in a wound. The problem with words, Deleuze claims, is that they are overburdened by habit, memory and history: 'their surface, barely broken, heals over again.'[142] The implication is that the total devastation—the 'absence of persistence, of revivification, or of regression'[143]—Malabou

[137] Laura Salisbury, '"What is the Word": Beckett's Aphasic Modernism', *Journal of Beckett Studies*, 17 (2008), 79–128, p. 79. Salisbury also engages with Luria's case history, contrasting Zasetsky's attempt to gain 'a victory over his aphasia … the rediscovery of a sense of self-cohesion, self-awareness and purposive subjectivity' with Beckett's less triumphant final text, p. 80.

[138] Luria, *Shattered*, p. 11.

[139] Luria, *Shattered*, p. 8.

[140] *The Letters of Samuel Beckett, vol. 1, 1929–1940*, ed. by Martha Dow Fehsenfeld and Lois More Overbook (Cambridge: Cambridge University Press, 2009), p. 518.

[141] Gilles Deleuze, 'The Exhausted', *Substance*, 24, 3 (1995), 3–28, p. 22.

[142] Deleuze, p. 22.

[143] Malabou, *Wounded*, p. 56.

claims is embodied in Beckett's writing was never achieved. His writing captures a tension between woundedness and healing, rather than embodying the unhealable wound itself. *The Man with a Shattered World* contains a similar tension between the habitual skin of the world and the wound that tears through its surface. Zasetsky's condition precluded him from participating in routine social activities: working, shopping, conversing, finding his way around. Once familiar words become strange and opaque; the surface does not heal. The impossible, the unreality of reality, thus erupts into view: 'words burst and backfire on themselves to reveal their own outside.'[144]

It is precisely these moments, in which routine existence is revealed as strange and arduous, that Beckett discerns in Marcel Proust's work: 'the perilous zones in the life of the individual, dangerous, precarious, mysterious and fertile, when for a moment the boredom of living is replaced by the suffering of being.'[145] *The Man with a Shattered World* inhabits such a perilous zone. Rebecca Comay characterises Proust's project as one of endless accumulation, which the author pursued with an obsessiveness bordering on the pathological. She compares his habit of scrupulously recording his experience of illness to Beckett's description of Molloy's farts:

> The involuntary processes of the body are registered in the archive of the voluntary memory until the very act of accounting is transformed into a kind of spiritual exercise—a running litany of complaints from which emerges at once both the possibility and the impossibility of writing.[146]

Comay invokes Walter Benjamin's discussion of the interaction of memory and forgetting in Proust which he likens to Penelope's shroud; every weaving is an unweaving.[147] In *The Origin of Tragic Drama* Benjamin similarly describes the search for origins as being animated by an

[144] Deleuze, p. 22. This is also strikingly similar to Viktor Shklovsky's analysis of Tolstoy's *War and Peace*. In his famous 1917 essay 'Art as Device', Shklovsky declares that Tolstoy's prose style 'estranges' objects in order to fully apprehend their significance: 'He [Tolstoy] does not call a thing by its name, that is, he describes it as if it were perceived for the first time.' Viktor Shklovsky, 'Art as Device' in *Theory of Prose*, trans. by Benjamin Sher (London: Dalkey, 1991), pp. 1–14, p. 6.

[145] Samuel Beckett, *Proust* (New York, NY: Grove Press, 1957), p. 8.

[146] Rebecca Comay, 'Proust's Remains', *October*, 144 (2013), 3–24.

[147] Benjamin, 'On the Image of Proust', *Selected Works, vol. 2, 1927–1930*, ed. by Michael W. Jennings, Howard Eland and Gary Smith (Cambridge, MA: Belknap Press of Harvard University Press, 1999), pp. 237–247, p. 238.

anachronous rhythm. For Benjamin, any discussion of origins must also contend with subsequent history: 'That which is original … needs to be recognised as a process of restoration and reestablishment, but, on the other hand, and precisely because of this, as something imperfect and incomplete.'[148] This dual movement also animates *The Man with a Shattered World*; though aimed at restoration the process always remains incomplete. Although Luria's case history functions as an attempt, performed by both doctor and patient, to tame the unruly profusion, obtrusion and confusion of words wrought by Zasetsky's injury, ultimately, it is not his recovery that determines his being, but his determination to recover:

> He was acutely aware of what it means to be human, and to the extent that his strength permitted, worked feverishly to overcome his problems. He suffered intensely, and though his world had been devastated, in the deepest sense he remained a man, struggling to regain what he had lost.[149]

Zasetsky goes in search of lost time, only to find its scattered and splintered remnants cannot be neatly reassembled; it is the persistence of their sharp edges and rough textures that define him.

Clark maintains that socialist realism was not only a literary genre but a form of cultural ritual. Zasetsky's attempts to describe his life might similarly be understood as ritualistic. It is a ritual that is repeated compulsively. The beginning is constantly returned to, the steps of the journey retraced: 'I've repeated the same points over and over again in my story and may do it again.'[150] Like the movement of the death drive which is always impelled to return to the beginning, Zasetsky pushes forwards by moving back through his most distant past: 'My memories come back from the wrong end … my recollections mostly have to do with my childhood.'[151] This movement is mirrored by the structure of the case history itself, which not only recounts the treatment in a non-linear fashion, but also includes a series of scientific 'digressions' (the Russian term *otstupleniye* also means 'retreat'). Although the narrative of the case history is progressive, it constantly loops back on itself: 'I'm still caught in a vicious circle. I can't break

[148] Benjamin, *The Origin of German Tragic Drama*, trans. by John Osborne (London: Verso, 2009), p. 45.

[149] Luria, *Shattered*, p. 35.

[150] Luria, *Shattered*, p. 80.

[151] Luria, *Shattered*, p. 96.

out of it.'[152] These reversals and repetitions nonetheless propel the narrative forwards. Returning to the original site of disaster and to the happier times that preceded it, also gestures towards the possibility of re-writing history or at least recovering something that was lost there: 'In the beginning it was all so simple'.[153]

In *Beyond the Pleasure Principle*, Freud, in a discussion of the compulsion to repeat, describes a child's game, which returns to the beginning over and over again. Fort/Da: the repetitive rhythm of the game represents an early attempt to gain mastery over the instincts. Benjamin suggested that Freud's description of repetition in play could also be understood as a 'reinstatement of the original condition'.[154] Politically, this implies the possibility of starting afresh, of renegotiating the course of events. This possibility is central to Comay's understanding of memory, which she also relates to the temporality of the death drive:

> Forgetting is not opposed to the work of remembrance. ... To forget to undo the past, to make it all "un-happen", is precisely to remember a moment before it all happened, to undo the inexorability of fate by re-staging the beginning, even if only in imagination and in proxy: to act *as if* we could take it over again, as if we could cast aside the legacy of dead generations, as if we could refuse the mourning work of cultural succession, as if we could cast off our patrimony, rewrite our origins, as if every moment, even those long vanished, could become a radically new beginning—unprecedented, unrehearsed, unremembered. This is why Freud will connect repetition with the death drive: the compulsion to repeat expresses a desire for inanimate existence and ultimately for non-existence: it is the desire to return to a time before the beginning—to go back not for the sake of regressing but in order to take it over, do it otherwise. The desire for repetition is essentially the desire for difference.[155]

This understanding of repetition indicates an understanding of recovery that is distinct from reintegration into an existing state of things. For Comay, Proust's drive to accumulate speaks of both anticipation and loss,

[152] Luria, *Shattered*, p. 158.

[153] Luria, *Shattered*, p. 3.

[154] Walter Benjamin, 'Toys and Play' (1928) in *Selected Writings*, vol. 2, pp. 117–121, p. 120.

[155] Rebecca Comay, *Mourning Sickness: Hegel and the French Revolution* (Stanford, CA: Stanford University Press, 2011), pp. 146–147.

it gestures towards the past as 'a constantly expanding repository of unrealised possibilities, botched initiatives, and thwarted hopes'.[156] The work of remembrance is therefore oriented towards the future; it returns to the past in order to recover and reactivate lost opportunities:

> It is rather about returning to a time before possibilities were eroded, to a moment before opportunities were squandered: what is being reactivated or repeated is not only the pressure of unrealised futures but of futurity itself in its ineffaceable obduracy—a potentiality all the more potent for being irretrievably past. Benjamin speaks of "hope in the past".[157]

In his furious writing and re-writing, weaving and unweaving Zasetsky returned again and again to the sites of squandered experiences, failure and lost chances, which were held in an irreconcilable tension with the violent actualities of subsequent history. It is precisely for this reason that *The Man with a Shattered World* emerges not only as a modernist document but also, obliquely, as a document of political hope.

Luria concluded by suggesting that Zasetsky 'reverted to the past' as a way of making sense of his experience in the present: 'he could not understand why the world had become so peculiar, why war was necessary, or find any justification for what happened to him.'[158] Zasetsky attempted to find a social justification for his condition, but he struggled to convince himself that his sacrifice had been worthwhile. Zasetsky seems to blame war for putting an end to the possibility for building a Communist society: 'were it not for war, the world would have become a great place to live long ago.'[159] He asks poignantly: 'What need is there for war, violence, slavery, oppression, murder, executions, poverty, hunger, back-breaking work, or unemployment in countries that have so much wealth?'[160] The Russian text explicitly says 'in capitalist countries' [*v stranakh kapitala*]

[156] Comay 'Proust's Remains', p. 5.

[157] Comay, 'Proust's Remains', p. 23. A similar argument animates Comay's work on Hegel: 'Possibility can appear only as a lost possibility, a lapsed possibility, a ruined possibility, even an impossibility, and the future as already passed.' Comay, *Mourning Sickness*, p. 3. Comay aligns Proust's work with Benjamin and reads them both as thinkers of the future. This reading differs from that proposed by Peter Szondi in his more conventional analysis of the differences between Benjamin and Proust. See, Peter Szondi, 'Hope in the Past: On Walter Benjamin', trans. by Harvey Mendelsohn, *Critical Inquiry*, 4, 3 (1978), 491–506.

[158] Luria, *Shattered*, p. 158.

[159] Luria, *Shattered*, p. 160.

[160] Luria, *Shattered*, p. 159.

but the mistranslation is suggestive nonetheless, as Zasetsky was coming to terms with a shattered Soviet world which had also failed to banish most of the social ills he lists.[161] Rather than looking to the officially ordained future of the socialist realist novel, *The Man with a Shattered World* returns to the past and in so doing opens up the possibility of thinking a different world. The case history documents Zasetsky's 'problems of coping with a world that seemed to have disintegrated'.[162] The problems he faced were particular to his neurological condition,[163] but the feeling of disaffiliation and disappointment the case history expresses has a social and political resonance that transcends Zasetsky's specific struggle. It is not insignificant that Luria's voice is sometimes indistinguishable from that of his patient.

The closing passages of the case history are overwhelmingly optimistic:

> In this age we have an opportunity to build and create a fine and beautiful world, to feed, clothe, and shelter all of mankind, not only the present generation but those for centuries to come. The water and earth of this world have an endless supply of energy and raw material, there is no need to fear any shortage of them. Soon there will be flights to outer space—first to the moon and the nearest planets. This will give us an even greater chance to enrich life with rare elements and substances that may be more plentiful on planets other than the earth.[164]

In the English text this bombastic utopian statement is undercut by the threat of human violence. The epilogue concludes with the sombre acknowledgement: 'We could do this, were it not for war.'[165] Zasetsky's confrontation with his own labyrinthine inner space undercuts the possibilities promised by visions of journeying into outer space. Both of these movements might be infinite, but instead of pushing outwards spatially into the unchartered cosmos, the text instead suggests that the future must be reached by way of a temporal movement back through history, which cannot banish the antagonisms of the past. The English version of

[161] Luria, *Romanticheskie Esse*, p. 233.

[162] Luria, *Shattered*, p. 51.

[163] Zasetsky notes mournfully that 'the average person will never understand the extent of my illness'. Luria, *Shattered*, p. 158.

[164] Luria, *Shattered*, p. 160. The Russian text contains a note explaining that Zasetsky had written this in 1957, two years before the first Soviet space mission (Luria, *Romanticheskie Esse*, p. 233).

[165] Luria, *Shattered*, p. 160.

the case history ends with an ellipsis rather a full stop, suggesting that Zasetsky's struggle is ongoing. The Russian text has a different ending, however. In the English version the authorial voices of Luria and Zasetsky merge, whereas in the Russian the closing statements reflecting on the future are attributed to Zasetsky, but the final sentence on war is omitted. Instead, Luria appends two concluding questions of his own: 'Need we say anything else? Is not this line, full of optimism, the best possible end to this short book?'[166] Again, by finishing with a question rather than a definitive statement, the ending remains open, but the Russian conclusion is more hopeful, suggesting that although the struggle to 'build and create a fine and beautiful world' may be continuous and arduous, it is not therefore futile or abandoned; 'we could do this'.

Monuments as Ruins: Between Modernism and Socialist Realism

Evgeny Dobrenko claims that one of the distinctive features of socialist realism was the 'elusiveness of its boundaries', arguing that socialist realist aesthetic forms were 'engrafted' onto Soviet consciousness with a thoroughness that gave shape to the Soviet experience and cannot therefore be confined to the realm of artistic production.[167] For Dobrenko, in the absence of real socialism, socialist realism was necessary to produce images of a transformed world. These images, he argues, eventually took on a reality of their own; socialism *was* socialist realism. That reality in the Soviet Union (whether it is understood to have been really socialist or not) was never identical to a socialist realist artwork and is of little concern to Dobrenko: 'The fact that everything is imagined in this "reality" only emphasises the status of *construction* and *image* of reality in Soviet culture, that is, the status of the aesthetic.'[168] This chapter has attempted to make a different claim: that socialist realist narratives (which themselves conformed to Marxist-Leninist understandings of historical development) permeated and gave shape to Soviet experiences and expectations (and were therefore not external to reality), but that experiences and expectations nonetheless contradicted and exceeded those narratives. *The Man*

[166] Luria, *Romanticheskie Esse*, p. 234.

[167] Evgeny Dobrenko, *Political Economy of Socialist Realism*, trans. by Jesse M. Savage (New Haven, CT: Yale University Press, 2007), p. xiv.

[168] Dobrenko, p. 14.

with a Shattered World registers discrepancies between socialist realism and Soviet reality rather than confirming their identity.

Unlike a socialist realist novel, Luria's case history was forced to contend with the reality of the irreparable. The drama of the socialist realist novel is driven by the hero's attempts to overcome obstacles and to sacrifice his private interests for the public good. Like Zasetsky, the ultimate goal of the socialist realist hero is social integration and the achievement of collective identity. As Hellbeck notes of Soviet diarists:

> The enlarged life of the collective was seen as the true subjecthood. It promised vitality, historical meaning, and moral value, and it was intensely desired. By contrast, a life lived outside the collective or the flow of history carried a danger of personal regression stemming from the inability to participate in the forward-thrusting life of the Soviet people.[169]

Kaganovsky discusses the 'trauma of returning home as unproductive and useless members of society', experienced by many veterans of the war.[170] Despite the presence of wounded heroes in Soviet fiction, disabled people, though numerous after the war, were not greeted as heroes but ostracised. Zasetsky's disability classification was downgraded in 1959, which exacerbated his recovery process by reducing his state benefits and declaring him fit to work.[171] Karl Levitin explicitly reflected on the gap between Zasetsky's struggle and the image of heroism routinely presented by the Soviet press who eagerly, 'report on a well-driller who saved a rig from blowing up, or a construction worker who covered a hole in a dam with his body', but would be unlikely to discuss Zasetsky's internal conquering of material nature in the same terms.[172] Zasetsky compared his own plight to the preoccupations of Soviet discourse in a slightly different manner, noting that 'there is a good deal of talk now about the cosmos and outer space', but

[169] Hellbeck, p. 10.

[170] Kaganovsky, p. 174. See also, Mark Edele, *Soviet Veterans of the Second World War: A Popular Movement in an Authoritarian Society, 1941–1991* (Oxford: Oxford University Press, 2008) and Elena Zubkova, *Russia After the War: Hopes, Illusions, and Disappointments, 1945–1957*, trans. by Hugh Ragsdale (Armonk, NY: M. E. Sharpe, 1998).

[171] K.E. Levitin, 'Confusion of Thought', *Journal of Russian and East European Psychology*, 36, 5 (1998), 20. On psychiatrists' roles in post-war rehabilitation of people wounded in the war, see Zajicek, pp. 168–227.

[172] Levitin, p. 7.

that interest in intergalactic flight detracted attention from a less grandiose kind of flight:

> The flight of a bullet, or a shell or a bomb fragment, that rips open a man's skull, splitting and burning the tissues of the brain, crippling his memory, sight, hearing, awareness—these days people don't find anything extraordinary in that.[173]

Esther Leslie blurs the neat line apparently separating modernism and realism, arguing that modernist art was not a rejection of reality, but a reaction to the new fractured realities of modernity:

> Modern experience is fragmentary. We experience the world in bits and pieces because we are alienated, because the world and ourselves are split in multiple ways. … Modernist art, in re-mediating that fragmentariness, produces a historically authentic mirror of experience. To that extent, it is an art of the real.[174]

In response to Malabou's question—'What mirror could reflect a brain?'—perhaps the mirror that could best reflect modern subjectivity is a broken one (and this goes against the more conventional understanding of 'reflection' espoused by Lenin and taken up by Luria). Leslie asserts that socialist realism 'closed the door on the exchange between realism and modernism, in an epoch when the divisions hardened'.[175] Despite the official silencing of these debates, however, these two apparently self-contained camps were always related to one another in complicated and contradictory ways. In the Soviet Union, the connection was quite literal: many of the artists and writers associated with the avant-garde were later forced to morph into realists of various kinds (as the example of Eisenstein, discussed in detail in the following chapter, demonstrates). Devin Fore has argued that the return of realism in Europe in the interwar years did not imply a regression. Realism after modernism could not be the same as realism before modernism; it bore the traces of the journey. Implicit in this thesis is the notion that realism

[173] Luria, *Shattered*, p. xxi.

[174] Esther Leslie, 'The Interrupted Dialogues of Realism and Modernism: "The Fact of New Forms of Life, Already Born and Active"' in *A Concise Companion to Realism*, ed. by Matthew Beaumont (Chichester: Wiley Blackwell, 2010), pp. 143–159, p. 144.

[175] Leslie, p. 156.

after modernism was still realism *during* modernity. Realism was forced to contend with a fundamentally transformed human subject: 'that body is no longer an organic whole and the human is no longer … a perfect integer.'[176]

The Man with a Shattered World is a portrait of a person who experienced the world in bits and pieces. He was alienated, split and incomplete. The process of writing was an attempt to reintegrate the subject back into a whole and to make him whole again in the process. Yet despite the hopeful assertions that litter the text, a moment of final reconciliation was never reached. Luria clearly stated that his patient could 'no longer synthesise'.[177] The scar that ran across Zasetsky's brain had an external counterpart in Soviet society; there was no reconciled whole into which Zasetsky could be integrated. The world itself was shattered. Zasetsky was a mutilated subject produced by the realities of modernity. In Luria's case history, conventional socialist realist forms were not sufficient for conveying the reality of a shattered world. In *The Man with a Shattered World* socialist realist and modernist literary forms, by necessity, converge.

According to Régine Robin, although the hero of the socialist realist novel can undergo tragedy, can end up alone, can experience failure, conflict and alienation, he cannot, she insists, be uncertain of his destiny, and the destiny of history with which he is in harmony:

> In the new man, there is necessarily and by definition and in the midst of disasters, mud and horror—a wager on the future, on the construction of a new society, a wager on the well-foundedness of the struggle that reconciles him at some point … with the movement of history, with which he turns out to be in harmony … there is the idea that the deficiencies will be overcome, and that the transparency of social and individual relations will finally be established.[178]

Zasetsky himself did not express doubt about the well-foundedness of the individual and collective struggle he continued to wage, but his optimistic proclamations were undergirded by a poignant acknowledgement that the catastrophes of war (and perhaps implicitly also the horrors that preceded

[176] Devin Fore, *Realism After Modernism: The Rehumanisation of Art and Literature* (Cambridge, MA, MIT Press, 2012), p. 4.

[177] Luria, *Shattered*, p. 28.

[178] Robin, pp. 242–243.

it) blocked the path to the version of the future that appeared glowing on the horizon in Soviet propaganda and socialist realist artworks.

Robin cautions against interpreting the official establishment of socialist realism as a straightforwardly oppressive top-down gesture, claiming that it emerged from heated collective debates that included thoughtful acknowledgements of the potential pitfalls of the genre and the dangers of 'inane optimism that no longer mirrors any aspect of the harsh struggles of the present'.[179] Despite all that was foreclosed in Soviet literature, Robin insists that ignoring the continued existence of polemic and variety, risks downplaying how much was still at stake in Soviet literature after the infamous 1934 Writer's Congress. She suggests that 'wrinkles on the surface of discourse … disclose the violence of the underlying aesthetic debate',[180] meaning that socialist realist novels never succeeded in presenting the future as a perfectly smooth surface. Attending to formal creases, folds and crinkles suggests that the image of reconciliation presented by the socialist realist novel, in tension with the historical present that produced it, always contained some disavowed acknowledgement of its own impossibility; an impossibility that comes more explicitly to the fore in Luria's non-fictional case history.

Although ostensibly staging a transition from spontaneity to consciousness, from the broken to the unified and from the wounded to the healed, *The Man with a Shattered World* is not, finally, a portrait of reconciliation. The real wounds of both subject and world cannot be ignored. However, Zasetsky's determination to heal and the case history's optimistic ending ensure that the work is emphatically not a portrait of resignation either. Rather than castigating it for falling prey to naïve optimism in a system that cannot be redeemed, perhaps it could instead be read an attempt to discover something in the contradictions, in the wrinkles; to find hope in the failures of the past. Robin imagines the majestic monuments of the fallen Communist regime coming to life and remembering their past. The monuments had attempted to instal a kind of forgetfulness, an uncomplicated image of wholeness. Now as they

[179] Robin, p. 18.

[180] Robin, p. 19. Comay, drawing on Proust and Benjamin, argues that wrinkles appear only after experiences have been lived. As such: 'the wrinkle is the perfect cipher of traumatic anachrony.' Comay, 'Proust's Remains', p. 24. She proposes a link between ageing flesh and the Proustian writing process: both register experience belatedly.

crumble, the contradictions they attempted to suppress emerge. Robin understands this process in psychoanalytic terms as a form of repression; the more the contradictions were denied, the more they threatened to erupt into view:

> Socialist realism which strove to block out opaqueness and misunderstanding, tragically destroyed the revolutionary social imaginary. I would say that the more fantasies of clarity, flamboyance, purity, homogeneity, transparence, fusion, and non-separation are nurtured, the more they repress obscurity, dissonance, gaps and holes, then the more obscurity, the unmasterable, dissonance, and events return into the rear. By driving out history, events precipitate themselves. By parrying distrust, madness precipitates itself. Fostering the romance only develops the horror.[181]

However, perhaps we can begin to recognise the monuments of the socialist regime as ruins even before they crumbled.[182] That is to say, the cracks were already present before the regime collapsed. Here we might recall Kracauer's claim that 'the position that an epoch occupies in the historical process can be determined more strikingly from an analysis of its inconspicuous surface-level expressions from that epoch's judgments about itself'.[183] The rough and wrinkled surface of Luria's text, saturated with familiar Soviet rhetoric yet riven with gaps and holes, registers the antagonisms of history. Indeed, it was Luria's development of a 'romantic' approach to science that allowed him to apprehend history. The text approaches the possibility of a transformed future via the past in the manner Comay describes as 'the fragile moment where the unforeseeable meets the immemorial—the point where sketch meets ruin'.[184]

[181] Robin, p. xxxvii.

[182] This is a reworking of Walter Benjamin's phrase: 'With the destabilising of the market economy, we begin to recognize the monuments of the bourgeoisie as ruins even before they have crumbled.' Walter Benjamin, *The Arcades Project*, ed. and trans. by Rolf Tiedemann (Cambridge, MA: Belknap Press, 1999), p. 13.

[183] Kracauer, 'The Mass Ornament' in *The Mass Ornament*, pp. 75–86, p. 75.

[184] Comay, *Mourning Sickness*, p. 147.

Conclusion

Theodor Adorno treats Beckett's *Endgame* as the exemplary work of art after the Second World War. Beckett succeeds in capturing the 'bombed-out consciousness'[185] that emerges from the debris, a smashed subjectivity that can no longer reflect on itself:

> After the Second World War everything is destroyed, even resurrected culture, without knowing it; humanity vegetates along, crawling, after events which even the survivors cannot really survive, on a pile of ruins which renders futile self-reflection of one's battered state.[186]

The Man with a Shattered World describes a wounded subjectivity akin to that described by Adorno. Luria's shattered subject bears a superficial resemblance to Malabou's 'new wounded'—humanity severed by the sudden violence of an unforeseen event. There is, however, a fundamental difference between Malabou's accident (which might be anything or nothing)[187] and Adorno's specific, unprecedented catastrophe. Ultimately Malabou's proposition is bleaker. She insists that the 'future harbours nothing to come',[188] whereas Adorno discerns a glimmer of hope in Beckett's work. As he comments in *Aesthetic Theory*: 'Even artworks that incorruptibly refuse celebration and consolation do not wipe out radiance.'[189] For Adorno, the terrain in *Endgame* is not empty, it is *almost* empty. An 'oblique light' shines.[190] Beckett specifies that the action plays out under a grey light.[191] When Clov looks through his telescope he sees 'Grey. [...] Grey! [...] GRREY!'.[192] The light is feeble, certainly, but it is not identical with darkness. In this near darkness 'consciousness begins to

[185] Adorno, 'Trying to Understand Endgame', trans. by Michael T. Jones, *New German Critique*, 26 (1982), 119–150, p. 123.

[186] Adorno, 'Trying...', p. 123.

[187] Malabou claims that the new wounded might emerge: 'as a result of serious trauma, or sometimes for no reason at all', *Ontology*, p. 1.

[188] Malabou, *Ontology*, p. 2.

[189] See also, Theodor Adorno, *Aesthetic Theory*, trans. by Robert Hullot-Kentor (London: Continuum, 1997), p. 82.

[190] Adorno, 'Trying...', p. 146.

[191] Samuel Beckett, 'Endgame' in *Samuel Beckett: The Complete Dramatic Works* (Chatham: Faber and Faber, 2006), pp. 89–134, p. 92.

[192] Beckett, 'Endgame', p. 107.

look its own demise in the eye, as if it wanted to survive the demise'.[193] A faint shard of hope glints dimly in the bleakness. And this tiny fragment—or 'smithereen' in Clov's terms[194]—is what constitutes the plot of *Endgame*: 'The tiny bit that is also everything—that would be the possibility that something could perhaps change.'[195] It is this tiny glimmer of possibility that Malabou fails to discern in Luria's text (and in Beckett's characters who she also aligns neatly with her cold and affectless 'new wounded'). Zasetsky may have been unable to fully recover but his fervent pursuit of recovery defined the narrative of the case history. Hope is maintained in the face of catastrophe. 'To think perhaps it won't all have been for nothing!'[196]

Zasetsky's sense of self derives from his struggle to heal rather than the achievement of healing. Luria's confrontation with the violence of history does not leave us in a position of resignation even if its conclusion is far from utopian. Although the path forwards is blocked, strewn with catastrophic debris, the journey to an improved future is never abandoned. The case history is dialectical, and the dialectic it describes, like the similarly compulsive death drive, constantly weaves and unweaves like Penelope: 'the work is infinite and the shroud forever unfinished.'[197] *The Man with a Shattered World* is animated by a negative dialectic, in which persistent conflicts impel history forwards. In the concluding passages of Luria's analysis of Zasetsky, he remarks:

> His wound healed twenty years ago, but the formation of scar tissue has resulted in attacks. The damaged areas of the cerebral cortex could not be restored. Hence, when he tried to think, his mind had to detour around these scorched areas and employ other faculties with which to learn and try to recover some lost skills. He desperately wanted to wake from this terrible dream, to break through the hopelessness of mental stagnation, to find the world clear and comprehensible instead of having to grope for every word he uttered. But it was impossible. ... He continues to try to recover what was irretrievable, to make something comprehensible out of all the bits and pieces that remain of his life.[198]

[193] Adorno, 'Trying...', p. 150.
[194] Beckett, 'Endgame', p. 97.
[195] Adorno, 'Trying...', p. 146.
[196] Beckett, 'Endgame', p. 108.
[197] Comay, *Mourning Sickness*, p. 130.
[198] Luria, *Shattered*, p. 158.

The 'bits and pieces' of the world can never become 'clear and comprehensible', but perhaps it is possible to make a 'detour around these scorched areas', not to a place of final reconciliation but at least engaging in a 'continuing struggle' and 'exhausting effort' in order to regain something of what has been lost, in the hope of creating something truly transformative in the future. Luria's case history seems to acknowledge that history is not driving towards a definite end point but that it might, nonetheless, be possible to carve out a path through the debris. Zasetsky was aware that he would never reach his destination, but the restless work of weaving and unweaving, must continue: 'He has returned to his story and is still working on it. It has no end.'[199]

[199] Luria, *Shattered*, p. 159.

CHAPTER 6

The Synaesthete

In letters to Luria, Oliver Sacks liked to cite T.S. Eliot's *Four Quartets*:

> And the end of all our exploring
> Will be to arrive where we started
> And know the place for the first time.[1]

Sacks echoed both Eliot's poetic lines and Luria's description of Zasetsky's continual struggle to heal—'it has no end'[2]—in a letter referring to the ongoing process of scientific research and experimentation: 'there is no end, only perpetual beginnings: every end is a beginning.'[3] This in turn recalls the movement traced in Sigmund Freud's *Beyond the Pleasure Principle* (1920) to which Luria had co-written an introduction at the beginning of his career:

> It is as though the life of the organism moved with a vacillating rhythm. One group of instincts rushes forward so as to reach the final aim of life as swiftly as possible; but when a particular stage in the advance has been reached, the other group jerks back to a certain point to make a fresh start and thus prolong the journey.[4]

[1] Sacks to Luria, June 7, 1975 and June 30, 1975, Oliver Sacks Archives, Oliver Sacks Foundation, New York City. See also: Oliver Sacks, *A Leg to Stand On* (London: Duckworth, 1984), p. 168.

[2] Luria, *Shattered*, p. 159.

[3] Sacks to Luria, June 30, 1975, Oliver Sacks Archives, Oliver Sacks Foundation, New York City.

[4] Sigmund Freud, 'Beyond the Pleasure Principle' in *The Standard Edition of the Complete Works of Sigmund Freud*, vol. 18, trans. and ed. by James Strachey et al. (London: Vintage, 2001), pp. 7–64, p. 41.

H. Proctor, *Psychologies in Revolution*, Mental Health in Historical Perspective, https://doi.org/10.1007/978-3-030-35028-4_6

This concluding chapter will similarly loop back to the start, returning to Luria's early career and retracing his conceptualisations of progression and regression, through a consideration of his engagements with and divergences from the Soviet film director Sergei Eisenstein (1898–1948), before ending (or beginning again) with a consideration of his 'romantic' case history *The Mind of a Mnemonist*, the synaesthete protagonist of which Eisenstein was also inspired by in his cinematic work.

Progress/Regress

'Highways, dams, railways. … The bustle of a big city. New machinery. New houses. New people'[5]—the planned epilogue to Sergei Eisenstein's unfinished film *Que Viva Mexico!* is set in a modern 'civilised' present of peace and prosperity. Funded by American socialists Upton and Mary Craig Sinclair, Eisenstein and his Soviet crew travelled in Mexico for a little over a year from December 1930 to February 1932 where they shot thousands of metres of film. Despite meditating on eternity and ancient traditions, Eisenstein's outlines for *Que Viva Mexico!* envision an oscillation between prehistory and modernity, death and life, stasis and dynamism; time hurtles forward inexorably. Masha Salazkina argues in her reading of the fragmentary remains of the unfinished project that although Eisenstein was increasingly drawn to moments of origin and unity, the imagined film was still animated by a dialectical movement: 'the return to the premodern is still mediated by its opposite, the ultramodern revolutionary impulse.'[6] Moving forward 'Towards Revolution. Towards a New Life',[7] *Que Viva Mexico!* was intended to begin in 'time remote'[8] and progress through distinct epochs of Mexican history until reaching its resolution in the future-oriented present.

In the middle of a notebook he bought in Aquila, which began in November 1931 and was kept during the last months of his travels through Mexico, Eisenstein pasted a letter from his friend Luria. Luria had recently returned from his first expedition to Soviet Central Asia where he had conducted experiments designed to demonstrate the cognitive impact of

[5] S.M. Eisenstein, *Que Viva Mexico!* (London: Vision, 1952/1971), p. 74.

[6] Masha Salazkina, *In Excess: Sergei Eisenstein's Mexico* (Chicago, IL: University of Chicago Press, 2009), p. 13.

[7] Eisenstein, *Que Viva Mexico!*, p. 73.

[8] Eisenstein, *Que Viva Mexico!*, p. 27.

the rapid social changes then taking place in the region, particularly the collectivisation of agriculture and the Campaign to Liquidate Illiteracy [*Likbez*] (discussed in the chapter titled 'The Primitive'). Indeed, Luria had written to Eisenstein in 1930 to invite the director to accompany his team on their trip.[9] With the same red pencil employed throughout the notebook to underline key phrases and words written in his own hand, Eisenstein highlighted the following passage in Luria's letter:

> Just think: what we assumed to be natural psychological processes (such as perception, comparison, association) are now revealed to be ***historical***, built in various stages of development! This is a big shift in psychology.[10]

Luria insisted that human perception and thought processes were socially determined and, like Eisenstein's representations of Mexico, he depicted the Uzbek Soviet Socialist Republic as a 'backward' culture entering modernity. Although Mexico and Soviet Uzbekistan both existed in 1931, Eisenstein and Luria understood those very different places in temporal terms and located them further back on a single developmental slope stretching from 'primitivism' to 'civilisation', the apotheosis of which would be a fully industrialised Communist society. Eisenstein's *Que Viva Mexico!* script outline makes clear that the social transformations he hoped to depict on screen was also intended to have a psychological counterpart, and he imagined a concluding moment of synthesis in which the vestiges of the past would be finally left behind:

> The same faces -
> but different people.
> A different country,
> A new, civilized nation.
> But, what is that?

With the victory of life over death and with the triumph of the present over the influences of the past, the film ends.[11] This envisaged conclusion, however, was never reached. Eisenstein was forced to abandon shooting

[9] Luria to Eisenstein, December 7, 1930, Moscow, The Russian State Archives of Literature and Art (RGALI), 1923-2-1130.

[10] Luria to Eisenstein, July 7, 1931, Moscow, The Russian State Archives of Literature and Art (RGALI), Sergei Eisenstein Papers, 1923-2-11301/2-69.

[11] Eisenstein, *Que Viva Mexico!*, p. 77.

before he had completed the final section of the film, which was to have depicted the Mexican Revolution of 1910. Focusing on the *soldaderas*, groups of women who accompanied the revolutionary armies, the film's 'fourth novel' would culminate in the birth of a child in the midst of death and defeat: an image of hope for the future. With this resolving part 'ruthlessly ripped out', Eisenstein declared that the unity and propulsive revolutionary thrust of his film had been destroyed.[12] Instead, the reams of unedited footage, drawings, photographs and notes generated by the project result in a heap of fragmentary materials in which past and present continue to coexist and overlap without auguring a transformed future; the intended 'victory' or dialectical overcoming was never realised.

Luria's letter to Eisenstein, which echoed the conclusion of the director's outline script, confidently imagined that a victory of the present over the past was being accomplished in Soviet Central Asia. But reading his letter alongside the disparate and chaotic materials in Eisenstein's notebook—which switched haphazardly between Russian, German, French and English, and included references to Wilhelm Reich, Dante, Shakespeare, G.K. Chesterton, Arthur Conan Doyle, Lucien Lévy-Bruhl and the Flemish mystic John of Ruusbroec alongside clippings of articles about weight loss, the psychic life of circus performers and popular detective stories—it is impossible to discern a through line emphasising themes of progress, synthesis or resolution. Although Luria and Eisenstein both set their sights on a classless future, for Eisenstein time jolted, looped, swirled or bounced forward whereas for Luria it was generally imagined to flow in a single direction occasionally punctuated by moments of crisis.

On 29 May 1932, just a few weeks after returning to the Soviet Union from his sojourn in Western Europe, America and Mexico, Eisenstein attended a late-night screening of his 1929 paean to collectivisation *The General Line* (released in Russia under the approved title *The Old and the New*) at the State Institute of Psychology in Moscow. The screening, hosted by Luria, was preceded by a lecture by the German Gestalt psychologist Kurt Koffka with Vygotsky translating from German for an audience of around 300 people. Koffka had been invited to the Soviet Union by Luria to join the second expedition to Central Asia which set off a few days later, on a long series of train journeys from Moscow to the

[12] Inga Karetnikova and Leon Steinmetz, *Mexico According to Eisenstein* (Albuquerque, NM: University of New Mexico Press, 1991), p. 3.

Uzbek city of Fergana.[13] Eisenstein embarked on the same journey in 1939 when he travelled to Soviet Uzbekistan to undertake a film celebrating the construction of the Great Fergana Canal, which, like *Que Viva Mexico!*, was planned to progress through successive epochs of the Republic's history (and was likewise eventually abandoned). Although Luria and Eisenstein both described history as advancing through specific stages, unlike Luria who emphasised the importance of 'removing the Muslim influence'[14] in Uzbekistan, Eisenstein drew aesthetic inspiration from Uzbek traditions, and hoped to show how aspects of the region's history resonated with and helped to shape the Communist present. As with his Mexican project, the journey to the future was accomplished through the intrusion of history into the contemporary moment. In an article intended for *Pravda* he proclaimed:

> The ancient madrasas of Samarkand, the ancient mosques of Bukhara, the ancient walls of Khiva have forever linked the pages of their ancient past to one of the most interesting pages in the history of the classless society in one-sixth part of the world. Because Bukhara and Samarkand, Khiva and Khorezm are the children of that same Uzbekistan in whose ring there sparkles its southern jewel, Fergana. That Fergana whose name will henceforth be inseparable from the history of the victories of Communism … how the real future will grow out of the depths of what is past, how a real understanding of the future will grow out of a profound knowledge of the past, how what is being created grows out of what has been created, just as how the theme of our tomorrow grows organically out of the past of this theme.[15]

Luria declared that people in Uzbekistan were making a 'leap of centuries'[16] to leave tradition behind, whereas for Eisenstein vestiges of the ancient might paradoxically enable the transition to the modern.

[13] See, Molly Harrower, *Kurt Koffka: An Unwitting Self-Portrait* (Gainesville, FL: University Presses of Florida, 1983), pp. 143–147.

[14] Luria, *Cognitive Development*, p. vi. As discussed in Chap. 2, Koffka explicitly denigrates Uzbek aesthetic judgment (a deficiency he ascribes to the Islamic ban on images) in a passage on Samarkand in Kurt Koffka, *Principles of Gestalt Psychology* (New York, NY: Harcourt, Brace and Company, 1935), pp. 349–350.

[15] RGALI 1923/2/919 cited and translated in Naum Kleiman, 'Fergana Canal and Tamburlaine's Tower', *Studies in Russian and Soviet Cinema*, 5, 11 (2011), p. 118. The article was published in an abridged form in *Pravda* on 13 August 1939 (which censored the text following the ellipsis in the passage cited here).

[16] Luria, *Cognitive Development*, p. 164.

In 'Socialist Senses: Film and the Creation of Soviet Subjectivity' (2012) Emma Widdis considers Soviet filmmakers' engagements with debates in Soviet psychology by asking how ideas 'filter down' from psychology to the arts.[17] Discussions of Eisenstein's relationship with Luria similarly tend to frame their interactions in terms of influence, which is generally assumed to flow in one direction—from psychologist to artist.[18] Yet as the vignettes sketched above indicate, even when they were engaged with similar subject matter or were in direct correspondence with one another, Luria's and Eisenstein's understandings of human perception and historical processes diverged as much as they converged. As such, this chapter will foreground the disciplinary ***distinctions*** between art and psychology that come to the fore when comparing these contemporaneous projects by juxtaposing moments and motifs from the two men's careers, setting contrasting elements 'into relations of conflict with one another', in a method inspired by Eisenstein's discussions of the montage technique.[19] Questions of temporality were central to both projects, which asked (among other things) how a Communist future might emerge from the present state of things; their answers, however, were often very different. Divergences between Eisenstein's and Luria's understandings of the relationship between the old and the new will be outlined, a tension between Luria's progressive model for understanding individual and historical development and Eisenstein's emphasis on non-linearity, shock and collision identified.

[17] Emma Widdis, 'Socialist Senses: Film and the Creation of Soviet Subjectivity', *Slavic Review*, 71, 3 (2012), 590–618, p. 595.

[18] On Eisenstein's relationship with Luria and Vygotsky, see, for example, David Bordwell, *The Cinema of Eisenstein* (Cambridge, MA: Harvard University Press, 1993), p. 176, Oksana Bulgakowa, 'From Expressive Movement to the "Basic Problem": The Vygotsky-Luria-Eisenstein Theory of Art' in *The Cambridge Handbook of Cultural-Historical Psychology*, ed. by Anton Yasnitsky, René van der Veer and Michel Ferrari (Cambridge: Cambridge University Press, 2015), pp. 423–448 and Julia Vassilieva, 'Eisenstein/Vygotsky/Luria's Project: Cinematic Thinking and the Integrative Science of Mind and Brain', *Screening the Past*, 38 (2013), http://www.screeningthepast.com/2013/12/eisenstein-vygotsky-luria%E2%80%99s-project-cinematic-thinking-and-the-integrative-science-of-mind-and-brain/#_edn15. It is significant that these analyses were all written by film scholars rather than historians of psychology.

[19] Sergei Eisenstein, 'Methods of Montage' in *Film Form: Essays in Film Theory*, trans. by Jay Leyda (San Diego, CA: Harvest Harcourt Brace Jovanovich Publishers, 1977), pp. 72–83, p. 78.

As a film-maker Eisenstein was interested in Luria's clinical work because it documented the ways in which 'abnormal' people perceived the world; his discussions with Luria and his patients and test subjects suggested ways of representing reality on screen that broke with conventional modes of perception. But as a psychologist Luria was dedicated to helping people overcome or manage their experiences so they could function in society. However, some of the people Luria encountered in his clinical work had permanent conditions; they would never become fully 'normal'. Luria may have consistently identified psychological harmony as a desirable goal (which had a counterpart in a Marxist-Leninist understanding of history's progression towards communism), but in practice his work attested to real experiences of fragmentation. As the previous chapter discussed, ameliorating the suffering of his patients could not simply involve assisting them to adjust to or correspond with dominant or 'healthy' modes of perception but necessitated approaching them on their own terms. This chapter's final section turns to a discussion of Luria's case history *The Mind of a Mnemonist*, the synaesthete protagonist of which Eisenstein evinced a particular fascination. In my concluding discussion I argue that in this case history, as in *The Man with a Shattered World*, Luria tempered the progressive frameworks that implicitly informed the majority of his published 'classical' writings. Despite acknowledging the material constraints faced by the people they depicted, I contend that these nuanced works ultimately provide a vision of individual and historical transformation that surpassed Eisenstein's by attending to struggles in the present without losing hope for the future.

Colliding Forms of Perception

In his discussion of Eisenstein's aesthetics, Peter Wollen declares that 'Eisenstein was influenced by two powerful, but in many ways incompatible teachers of psychology: Freud and Pavlov'.[20] Eisenstein's connection with Luria, which Wollen overlooks, was more concrete. Luria had attended Eisenstein's production of *The Wiseman* in 1923 soon after moving to Moscow and the pair remained in touch until the director's

[20] Peter Wollen, *Signs and Meaning in the Cinema* (London: Secker & Warburg, 1969), p. 36.

death 25 years later.[21] Wollen emphasises Pavlov's influence on Eisenstein above all other psychologists, but neglects to consider Eisenstein's personal connections with either non-Pavlovian Soviet psychologists like Luria and Vygotsky or the German Gestalt psychologists to whom Luria introduced Eisenstein in the late 1920s.[22] Eisenstein described Luria as his 'unfailing adviser, friend and consultant'[23] and credited the psychologist with introducing him to 'a large number of secrets of experimental psychology, usually inaccessible to ordinary mortals'.[24] Not only did Luria and Eisenstein discuss psychological and aesthetic questions, but they also had many overlapping interests: in psychoanalysis, in primitivism, in childhood and in pathology.

Both Eisenstein and Luria lived and worked through the heady post-revolutionary period and despite experiencing criticism and censorship, survived the Stalinist purges and continued to work through the Second World War. Their careers existed in fraught relationship to the Soviet state and its erratic yet prescriptive vocabularies and policies. Euphemistic references to the constraints of the period pepper the letters that Luria sent Eisenstein throughout their acquaintance, subtly bewailing, for instance, the difficulties they experienced in obtaining permission to travel abroad.[25] Analyses of their respective oeuvres necessarily contend with the fraught question of their relationships to the regime and ask to what extent their visions were compromised by Stalinist imperatives, often reaching very different conclusions.[26] Scholars have read Eisenstein's final film *Ivan the*

[21] See Oksana Bulgakowa, *Sergej Eisenstein: Eine Biographie* [*Sergei Eisenstein: A Biography*] (Berlin: Potemkin Press, 1997), p. 80, p. 112. Elena Luria claims that they first met in 1925 or 1926, shortly after the release of *Battleship Potemkin*. Elena Luria, *Moi Otets: A.R. Luria* [*My Father: A.R. Luria*] (Moscow: Gnosis, 1994), p. 121.

[22] In addition to the contact with Koffka, Luria acted as a go-between, putting Eisenstein in touch with Kurt Lewin in Berlin. See, Luria to Kurt Lewin, 20 September 1929, 22 November 1929. Akron, OH, Archives of the History of American Psychology (AHAP), Kurt Lewin Papers, Box M2931, Folder 2.

[23] S.M. Eisenstein, 'On Folklore', in *Selected Works: Volume 4, Beyond the Stars: The Memoirs of Sergei Eisenstein*, ed. by Richard Taylor, trans. by William Powell (London: BFI Publishing, 1995), pp. 592–615, p. 611.

[24] Eisenstein also mentions that he began a collaborative seminar with Luria, Vygotsky and the linguist Nikolai Marr, exploring the 'problems of the nascent language of cinematography'. See Eisenstein, *Metod: Grundproblem* [*Method: Grundproblem*] (Moscow: Muzey Kino Eizenshtein Tsentr, 2002), p. 136.

[25] Alexander Luria to Sergei Eisenstein, February 17, 1930, RGALI 1923-1-19321/26.

[26] For a particularly crude example that argues Eisenstein's engagement with the Soviet state invalidated his aesthetic approach from the beginning, see Ron Briley, 'The Artist in Service of the Revolution', *The History Teacher*, 29, 4 (1996), pp. 525–536.

Terrible, Part II (completed in 1946 but banned until 1958) as both a justification and a critique of Stalinism.[27] Similarly, Luria has been characterised as both complicit with the brutal regime and as its victim.[28] These conflicting accounts indicate that Eisenstein's and Luria's careers resist neat characterisation: both men produced work under duress which necessarily worked within the constraints of a particular idiom without ever being reducible to it. Indeed, the differences between Eisenstein's and Luria's understandings of Marxism, communism and the unfolding of history discussed in this chapter indicate the extent to which discrepancies in thought persisted even during the most oppressive periods of Soviet history.

Luria granted Eisenstein access to his patients and laboratories. Luria in turn made use of the Institute of Cinematography to produce scientific instructional films.[29] In *The Cinema of Eisenstein*, David Bordwell claims that the director's 'discussion of prelogical thought reflects his continuing affiliation with Vygotsky and Luria, who studied children, preliterate cultures and brain-damaged patients in search of cognitive processes at variance with deductive reasoning'.[30] Eisenstein also engaged with Freud's *Beyond the Pleasure Principle*, pondering the implications of the death drive for art, which Luria and Vygotsky had written the introduction to when it appeared in Russian translation in 1925.[31] As Bordwell observes, Luria was indeed interested in the relation between the ontogenetic and the phylogenetic. His work did indeed also dwell on moments of origin and processes of development. However, unlike Eisenstein, Luria had little interest in celebrating the thought processes of children, preliterate cultures and brain-damaged patients.

[27] See Kristin Thompson, '"Ivan the Terrible" and Stalinist Russia: A Reexamination', *Cinema Journal*, 17, 1 (1977), pp. 30–43.

[28] For an example of an article arguing for Luria's complicity with Stalinism, see Uwe Gielen and Samvel Jeshmaridian, 'Lev Vygotsky: The Man and the Era', *International Journal of Group Tensions*, 28, 3/4 (1999), pp. 273–301. For an account that emphasises victimisation, see Evgenia D. Homskaya, *Alexander Romanovich Luria: A Scientific Biography*, trans. by Daria Krotova (New York, NY: Plenum Press, 2001).

[29] In letters to Kurt Lewin Luria mentions a film made in association with the Film Institute on counting and attention in children. See Luria to Lewin, 22 November 1929 and 12 December 1929. Akron, OH, Archives of the History of American Psychology (AHAP), Kurt Lewin Papers, Box M2931, Folder 2. This collaboration with the Film Institute is also mentioned in Homskaya, p. 39.

[30] David Bordwell, *The Cinema of Eisenstein* (Cambridge, MA: Harvard University Press, 1993), p. 176. Masha Salazkina similarly claims that Eisenstein's interest in 'primitive' forms of thought was influenced by Luria. See Salazkina, p. 184.

[31] See Anne Nesbet, *Savage Junctures: Sergei Eisenstein and the Shape of Thinking* (London: I.B. Tauris, 2003), p. 201.

Eisenstein declared: 'Let's work ... backwards'.[32] Luria, on the other hand, faced forward. He studied supposedly 'backward' figures in the hope that they might shed light on and help to precipitate a transition to an ideal form of 'advanced' cognition. The children, 'primitives' and brain-injured people he encountered in his clinical work were thus understood in relation to an ideal norm from whom they departed: the 'civilised' educated adult. Luria's progressive understanding of individual development could be understood as a counterpart to an orthodox Marxist-Leninist understanding of historical development. He built on Vygotsky's theoretical writings which conceptualised human life as an 'ascent up the stages of development', a movement he explicitly likened to the movement of history.[33] Luria sought to trace and even encourage the settling of 'chaotic' thought processes into 'orderly' ones. In practice this typically meant demonstrating that a person comprehended shared social structures and concepts—praising, for example, the child who could define a trade union or *kolkhoz* (collective farm)—a capacity he linked to the development of abstract thinking.[34] He thus viewed the past as something that could and should be left behind: civilisation succeeded primitivism, adulthood succeeded childhood.

Eisenstein's concerns were similarly oriented towards a harmonious future, and the example of his unfinished Mexican project suggests that he shared some of Luria's progressive assumptions. However, his work tended to prioritise conflict, shock and superimposition, rather than depicting individual and historical progress as a journey through a successive series of ascending stages. Even in the outlines for *Que Viva Mexico!* which are ordered chronologically and extol 'modernisation', Eisenstein suggested that history might collide with the present to create something new, whereas Luria was concerned with overcoming the past. As Luria noted in an essay on childhood development, as it grows up a child 'loses or discards all the earlier, more primitive forms ... each subsequent stage super-

[32] Eisenstein, 'The Psychology of Art' in *The Psychology of Composition*, ed. and trans. by Alan Unchurch (London: Methuen, 1988), pp. 1–15, p. 8.

[33] Vygotsky, 'The Problem of Age', (1932–1934) in *The Collected Works of L.S. Vygotsky*, vol. 5, trans. by Marie J. Hall, ed. by Robert W. Riener (New York, NY: Plenum Press, 1998), pp. 187–205, p. 188.

[34] Luria, 'Speech and Intellect among Rural, Urban and Homeless Children' in *Soviet Developmental Psychology*, ed. by Michael Cole (New York, NY: M. E. Sharpe, 1977), pp. 32–64.

sedes and supplants the previous one'.[35] Eisenstein hoped to overturn established modes of perception, whereas Luria sought to assimilate or reintegrate people into pre-existing orders.

This difference in emphasis is evident in Eisenstein's essay 'Beyond the Shot' (1929). Eisenstein justified his abandonment of realist proportion and scale in cinema through a discussion of children's drawings shown to him by Luria. In these sketches the children have not obeyed the laws of realism but depict the object of most importance as much larger than other objects in the scene. The proportions of the image correspond to the relative significance of the different objects, rather than their literal size. Eisenstein used this example to launch an attack on 'orthodox formal logic' which attempts to uphold 'the inviolable order of things'. He announced that 'positivist realism is by no means the correct form of perception. It is simply a function of a particular form of social structure, following on from an autocratic state uniformity of thought.'[36] The principle of montage creates 'collision' and 'conflict', blasting through the staleness and 'ideological uniformity' of received modes of perception.[37] Luria's psychological work was mobilised by Eisenstein in support of an art form that would unsettle people's habitual ways of seeing.

Eisenstein suggested that children point to ways of disrupting the structures of the adult world, whereas Luria sought to instil in children an understanding of representation in accordance with those established forms of logic. Luria discussed the same drawing referenced by Eisenstein (of a room in which a box of matches is presented as disproportionately large) in 'The Child and its Behaviour' (first published in 1930). However, rather than praising the child's unique understanding of perspective, he points instead to the child's failure to perceive the connections between things:

> A complex systematic picture of the world, an arrangement of its phenomena according to their connections and causative relations is replaced by a simple 'glueing' together, a primitive combining, of isolated features.[38]

[35] Luria, 'Experimental Psychology and Child Development' in *The Selected Writings of A.R. Luria*, ed. by Michael Cole (White Plains, NY: M. E. Sharpe, 1978), pp. 78–96, pp. 94–95.

[36] Eisenstein, 'Beyond the Shot' in *Volume 1, Selected Writings, 1922–1934*, ed. by Richard Taylor (London: BFI, 1988), pp. 138–150, p. 142.

[37] Eisenstein, 'Beyond the Shot', p. 144.

[38] Luria and Vygotsky, 'The Child and Its Behavior' in *Studies on the History of Behavior: Ape, Primitive and Child*, ed. and trans. by Victor I. Golod and Jane E. Knox (Hillsdale, NJ: Lawrence Erlbaum, 1993), pp. 140–231, p. 157. According to the book's introduction 'The Child and Its Behavior' was written solely by Luria. See 'Translator's Introduction', pp. 1–35, p. 24.

In the same discussion Luria also includes a drawing of a figure on a horse by an illiterate Uzbek peasant, a reproduction of which is preserved in the Eisenstein archives. Luria links childhood perception to this 'primitive' representation, claiming that both isolate elements 'without synthesising them into an integral image'.[39] In highlighting the shortcomings of childhood perception and the perception of an 'Uzbek woman with little cultural background',[40] Luria's position seemed closer to one criticised by Eisenstein than to Eisenstein's own pronouncements in that he could be accused of '[d]isplacing the expressiveness of archaic disproportion for regulated "stone tables" of officially decreed harmony'.[41]

Beginnings and Endings

When Eisenstein died in 1948 Luria is said to have performed an autopsy, slicing open his friend's skull to search for material signs of creative genius. Luria kept photographs of the film director's brain until his own death 30 years later, showing his students the unusually enlarged right hemisphere (the section of the brain responsible for visual processes).[42] Other accounts claim Luria preserved the brain itself in a glass vessel filled with alcohol 'because it had an interesting shape'.[43] In life, Eisenstein had his own collection of deathly objects. Nestled among the books, antiquities and Mexican handicrafts in his Moscow apartment, he kept a human foetus in a jar of formaldehyde. He was moved to acquire the specimen, which floated eerily on the threshold between prehuman and human, after experiencing a foetus die in his hands during a shoot for an educational film about abortion in Switzerland.[44] As his own life neared its end, Eisenstein

[39] Luria, 'The Child and Its Behavior', p. 158.

[40] Luria, 'The Child and Its Behavior', p. 158.

[41] Eisenstein, 'The Cinematographic Principle and the Ideogram' in *Film Form*, pp. 28–44, p. 35. This essay (also from 1929) contains a discussion of the children's pictures shown to Eisenstein by Luria similar to that discussed in 'Beyond the Shot'.

[42] V. Ivanov, *Chet i nechet: Asimmetriia Mozga i Znakovykh Sistem* [*Odd and Even: Asymmetry of the Brain and Sign Systems*] (Moscow: Sovetskoe Radio, 1978) cited in Julia Vassileva, 'Eisenstein/Vygotsky/Luria's Project: Cinematic Thinking and the Integrative Science of Mind and Brain', *Screening the Past*, 38 (2013).

[43] K.E. Levitin, 'Epilogue: Luria's Psychological Symphony', *Journal of Russian and East European Psychology*, 6, 36 (1998), pp. 33–62. Luria's great-granddaughter claims that Eisenstein's brain was stored in Luria's fridge at home prior to dissection. See 'Dachi', *Afisha*, July 26, 2011, https://daily.afisha.ru/archive/gorod/archive/summerhouse-photostory/.

[44] This episode is recounted in Nesbet, p. 140.

became increasingly preoccupied with beginnings and endings. On the ontogenetic level, he dwelled on intrauterine experience and death. He also returned to the beginnings of his own life, composing a memoir reflecting on his formative childhood impressions and experiences. In this autobiographical work he noted that his 'life had passed at a gallop, without a backward glance, in constant transit, leaving one train to catch after another', but that he was now inspired to pause and cast his eyes back over the journey.[45] These interests had a phylogenetic counterpart: Eisenstein's writings turned to explorations of the primitive, prehistoric and primordial.[46]

On the 19 September 1947, just five months before his death, Eisenstein wrote in his diary:

> Alexander R Luria telephoned me yesterday, inviting me to give a series of lectures on the psychology of art at the Psychology Institute of Moscow University (for students of the senior class). Not having made up my mind to do it, I naturally started planning this morning while in bed how I would present such a series.[47]

In the conspectus he submitted to Luria for their planned collaborative lecture course on 'The Psychology of the Creative Process', Eisenstein outlined his intention to discuss art through an analysis of life in its embryonic form and an exploration of the sensuousness of thought prior to the acquisition of language. This chimes with his discussion of human psychology in his autobiography:

[45] Eisenstein, *Immoral Memories: An Autobiography*, trans. by Herbert Marshall (London: Peter Owen, 1985), p. 3.

[46] Nesbet describes this shift in Eisenstein's thought away from shock and montage (as discussed by Wollen) in the following terms: 'Gradually Eisenstein was shifting away from his earlier emphasis on the abrupt, violent changes that shifts the course of history, towards an interest in the early protoplasmic forms of life, in the shapes and forms of things before identity is determined.' p. 142. On Eisenstein's late interest in early forms of life, see also Esther Leslie, *Hollywood Flatlands: Animation, Critical Theory and the Avant-Garde* (London: Verso, 2002), pp. 222–237, Yon Barna, Eisenstein (Bloomington, IN: University of Indiana Press, 1973), p. 268 and Salazkina, p. 133.

[47] Eisenstein, 'Conspectus of Lectures on The Psychology of Art' in *The Psychology of Composition*, ed. and trans. by Alan Upchurch (London: Methuen, 1967), pp. 16–25 (transliteration modified for consistency).

> Regarding 'man' again, it is interesting that the presence of the human (very human!) basis interests me in its most 'not yet human' stage. That is, in all those spheres and beginnings where man is present in art in a way that is hidden, not yet manifest.[48]

He also retroactively characterised his trip to Mexico as responding to this interest in origins, describing it as a journey back in time 'to the sources of human culture'.[49] His notes for the conspectus suggest that he saw the artwork as a means of accessing and re-experiencing earlier stages of existence: 'Plunging "back" to the stage of sensuous thought, we lessen partial control.'[50]

In his 1946 essay 'On Folklore', Eisenstein described how he increasingly became interested in 'worlds which are dormant within us'.[51] He dated this shift in his thinking to a 1935 speech delivered at the All-Union Creative Conference of Workers in Soviet Cinematography, in which he controversially declared that artworks needed to reawaken ancient modes of perception: 'Art is nothing else but an artificial retrogression in the field of psychology towards the forms of earlier thought processes.'[52] He also recalled a significant discussion with Luria, which he credited with provoking his initial interest in the subject of biological origins.[53]

Like Freud in *Beyond the Pleasure Principle*, Eisenstein suggested that life was bookended by moments of reconciliation. In the course outline he drew up for Luria in 1947, he noted: 'In the embryo, there is not yet a conflicting reaction.'[54] Eisenstein, like Freud, imagined a quiescent state from which life emerged and to which it would eventually return.[55] Eisenstein's return to early forms of life, civilisation and thought was still conceived of in revolutionary terms, however. He claimed to be drawn to

[48] Eisenstein, *Immoral Memories*, p. 217.

[49] Eisenstein, *Immoral Memories*, p. 212.

[50] Eisenstein, 'Conspectus', p. 25.

[51] Eisenstein, 'On Folklore', p. 611.

[52] Eisenstein, 'Film Form: New Problems' in *Film Form*, pp. 122–149, p. 144. Alongside a long discussion of primitive cultures Eisenstein also mentions having been introduced to people being treated for brain injuries in Moscow who exhibit similar patterns of thought, presumably with Luria as his guide, p. 141.

[53] Eisenstein, 'On Folklore', p. 611.

[54] Eisenstein, 'Conspectus', p. 19.

[55] Eisenstein also drew on psychoanalytic discussions of conception and birth. For example, in 'On Folklore' he mentions being introduced to Sandor Ferenczi's work on gentility by Hanns Sachs, p. 600.

the earliest stages of life and civilisation because they, like the imagined Communist future, were classless: 'To where we are plunging them back. To paradise. To the stage of non-differentiating thought. But also the pre-class stage. And therein lies the fascination.'[56] Plunging back was therefore also conceived of as a pushing forward. As Anne Nesbet discusses, for Eisenstein

> prehistory is never exactly left behind, but rather turned into a dialectical component of the present historical or aesthetic moment: even the highest most developed image bears at least a plastic trace of its most primitive ancestry.[57]

He hoped that exploring birth, the dawn of humanity and the beginnings of life itself might dialectically combine with the contemporary through cinema to create an ecstatic feeling in his audience. Eisenstein's later work does not dispense with the dialectic, but a collision of objects in space (both on screen and in the cinema auditorium) was replaced with a collision of moments in time.

Eisenstein, like Luria, aligned the thought processes of children with those of primitive peoples; he aligned individual with historical and even with evolutionary development. He too framed this as a transition from figurative to abstract thought: 'The individual progressively moves from thinking in images and emotions to the maturity of consciousness.'[58] Unlike Luria, however, whose work tended to describe how these early stages would ideally be surpassed, Eisenstein hoped that a great artwork could access dormant yet still extant modes of perception. This involved going beyond the mechanistic physiology of conditioned reflexes that dominated his earlier work (as discussed by Wollen) to access a more distant organic substrate of being:[59]

[56] Eisenstein, 'Conspectus', pp. 16–25.

[57] Nesbet, p. 11. Nesbet argues that the vacillating movement that animates Freud's *Beyond the Pleasure Principle* also courses through Eisenstein's later work, observing an oscillation between life and death or womb and tomb in *Ivan the Terrible, Part I*, pp. 198–208.

[58] Eisenstein, 'On Folklore', p. 602.

[59] Interestingly, Bulgakowa argues that Eisenstein's shift in thought was inspired by Vygotsky: 'Vygotsky's interpretation of the dual structure of aesthetic reaction is precisely what spurred Eisenstein to abandon the reflexological model of montage of attractions in favour of understanding works of art as dynamic unities of opposites that both model and provoke a cathartic response.' Bulgakowa, 'From Expressive Movement to the "basic prob-

> A responsive reaction is triggered deep within me—beyond the convulsions of the brain, somewhere in the tissue—by the very structures which are my contemporaries from the time when I was an individual on the evolutionary ladder, no more than a child; an embryo, a ball of albuminous protoplasm or a fertile drop, which is all that I once had been.[60]

Eisenstein insisted that this regressive movement was progressive. The process of digging down to reach the most primal forms of thought was addressed to transforming the present. Ancient elements would be brought to the surface where they would combine with the most advanced forms of consciousness to take on 'new, ultra-contemporary forms'.[61] According to Eisenstein, this dual movement was integral to great artworks, describing how 'a determined progressive ascent towards ideas at the highest peak of consciousness' must be combined with 'a penetration through the structure of form into the deepest layer of emotional thinking'.[62]

The developmental model that underpinned Luria's psychological theories—from his early rejection of the Freudian death drive,[63] through his work with illiterate people in Uzbekistan and experiments with children—does implicitly posit a moment of final unity and reconciliation akin to the classless harmony described by Eisenstein. He did not, however, suggest that this recapitulated a lost earlier form on a new level. Unity is only located in the future; Luria does not look back. As such, his work is devoid

lem"', p. 435. Bulgakowa claims that Eisenstein was given a manuscript copy of Vygotsky's *Psychology of Art* in 1925, a text whose influence on Eisenstein is also discussed in Karla Oeler, 'A Collective Interior Monologue: Sergei Parajanov and Eisenstein's Joyce-Inspired Vision of Cinema', *The Modern Language Review*, 101, 2 (2006), 472–487. There is no comparable text (reflecting explicitly on art and aesthetics) among Luria's many publications.

[60] Eisenstein, 'On Folklore', p. 603.

[61] Eisenstein, 'On Folklore', p. 605.

[62] Eisenstein, 'Speeches to the All-Union Creative Conference of Soviet Filmworkers' in *Selected Works: Volume 3: Writings, 1934–1947*, ed. by Richard Taylor, trans. by William Powell (London: BFI Publishing, 1996), pp. 16–46, p. 38.

[63] Luria and Vygotsky's introduction to the Russian translation of *Beyond the Pleasure Principle* (published in 1925) rejected the 'pessimism' of Freud's essay and instead argued that a transformation of social conditions could transform individual consciousness in turn. See Alexander Luria and Lev Vygotsky, 'Introduction to Beyond the Pleasure Principle' in *The Vygotsky Reader*, ed. by René van der Veer and Jaan Valsiner (Cambridge, MA: Blackwell, 1994), pp. 10–18. For a detailed discussion of their reading of Freud, see Hannah Proctor, '"A Country Beyond the Pleasure Principle": Alexander Luria, Death Drive and Dialectic in Soviet Russia, 1917–1930', *Psychoanalysis and History*, 18, 2 (2016), 155–182.

of the stark opposition of opposing forces integral to Eisenstein's theories. Despite noting his increased interest in early forms of life, Eisenstein did, however, identify a unifying theme that ran throughout his written and cinematic oeuvre:

> If I were an impartial researcher, I would say of myself: this author appears to be constantly fixated on one idea, one theme, one subject. Everything he has thought up and done, not only within the different films but through all his plans and films, is in each and every case one and the same thing. Almost invariably, the author uses different periods … different countries and peoples … different social movements and processes within the shift towards different social forms, as different masks covering one and the same face. This face is the realisation of the ultimate goal—the attainment of unity.[64]

Luria may not have been interested in recapturing original unities in this manner but his future-oriented work was similarly concerned with an ultimate goal: the 'advanced' human subject (and his or her implicit historical counterpart: civilisation or communism).

However, towards the end of his career Luria, contra Eisenstein, moved away from an emphasis on synthesis to an acknowledgment of fragmentation in his 'romantic' case histories; his career thus followed an inverse trajectory to Eisenstein's. In this chapter's opening section I noted a contrast between Eisenstein's Mexican film script outline and the chaotic notes generated by the trip. A similar contrast might be observed in Luria's work between his ideal frameworks for understanding human development and the actual people he encountered in his research, who often exceeded or deviated from the progressive trajectories his developmental theories attempted to sketch out. Luria's two literary case histories, written towards the end of his long working life, not only mark a break from this tendency but also point to a more nuanced, sombre and conflicted political vision than Eisenstein's eschatological equation of a classless society with intrauterine experience or primal protoplasm, not to mention a more materialist one. The junctures, collisions and shocks that characterised Eisenstein's films of the 1920s find an unexpected counterpart in Luria's case histories—*The Mind of a Mnemonist: A Little Book about a Vast Memory* (1968) and *The Man with a Shattered World: The History of a Brain Wound* (1971)—which describe pathological conditions that preclude their subjects from assimilating into existing social structures.

[64] Eisenstein, *Selected Works, Volume 4*, 'The Author and his Theme', pp. 787–796, p. 792.

Luria characterised his case histories as belonging to a lost tradition of 'romantic science' and contrasted them to the 'classical' approach typical of most scientific writing, his own works included. As Luria noted in his first emphasis-laden letter to Oliver Sacks: 'it is a kind of "Romantic Science" which I wanted to introduce, partly because I am strongly against formal statistical approach and for a qualitative study of personality, for every attempt to find factors underlying the structure of personality.'[65] For Luria, the romantic scientist should convey the qualitative specificity of human experience by abjuring abstraction:

> Romantics in science want neither to split living reality into its elementary components nor to represent the wealth of life's concrete events in abstract models that lose the properties of the phenomena themselves. It is of the utmost importance to romantics to preserve the wealth of living reality, and they aspire to a science that retains this richness.[66]

In these case histories, Luria was not concerned with describing symptoms in isolation from a person's whole personality, but hoped that his granular style and empathetic approach would allow for the preservation of 'the manifold richness of the subject'.[67] *The Mind of a Mnemonist* opens with an epigraph from Lewis Carroll's *Through the Looking-Glass* (a text of which Eisenstein was also fond) and Luria declared his intention to take his readers through the 'smooth, cold' glass to explore the unfamiliar world on the other side.[68] As was evident in the texts discussed in the first three chapters of this book, the majority of Luria's publications described people in relation to abstract ideals, whereas his 'romantic' works sought to understand the people they described on their own terms. Sacks noted in his correspondence with Luria that 'you couldn't write as you do unless you knew your patients from "inside"; an intense exercise of empathy until you almost become the *Mnemonist* or *The Shattered Man*.'[69] Eisenstein's early films were famously criticised in the Soviet Union for alienating 'ordinary' Soviet people. Indeed, he enlisted Luria's and Vygotsky's help in drawing up a questionnaire for audiences of his films to establish whether they had interpreted the symbolic visual images and montages as

[65] Luria to Sacks, July 19, 1973, Oliver Sacks Archives, Oliver Sacks Foundation, New York City.

[66] Luria, *The Making of Mind*, p. 174.

[67] Luria, *The Making of Mind*, p. 178.

[68] Luria, *The Mind of a Mnemonist: A Little Book about a Vast Memory*, trans. by Lynn Solotaroff (London: Penguin Books, 1975), p. 5.

[69] Sacks to Luria, June 14, 1975, Oliver Sacks Archives, Oliver Sacks Foundation, New York City.

the director intended.[70] Luria's works demonstrate that many people already perceived the 'ordinary' world as shattered, splintered or swirling; Eisenstein's formal innovations may therefore have been more realist (or at least more realistic) than many commentators perceived.

As discussed in the previous chapter, *The Man with a Shattered World* described a patient who survived a bullet wound to the head that severely altered his cognitive capacities. Luria tacitly acknowledged that the violence of the Second World War rendered a neat linear developmental framework for understanding individual historical development untenable; in a shattered world total healing is impossible. Eisenstein's early experiments in cinematic form were intended to find an aesthetic equivalent to revolution, whereas Luria's turn from unity to fragmentation was a response to the violent impact of war. Confronted with a patient who could not be cured but who nonetheless retained a determination to improve his condition, negativity made an incursion into Luria's work. But it was the protagonist of Luria's earlier case history that fascinated Eisenstein.

Suffering from Reminiscences

Intrigued by Luria's accounts of a synaesthetic patient with an extraordinary memory, Eisenstein asked that the psychologist introduce them.[71] In his essay 'On Color' (1937–1940), Eisenstein discussed how his reflections on the use of colour in cinema were inspired by his encounters with Luria's synaesthetic patient, Solomon Shereshevsky, about whom Luria eventually wrote in *Mind of a Mnemonist*.[72] Eisenstein declared that colour is the solution to finding synchronicity between sound and image, but he insisted

[70] Michael Cole relayed this in conversation with Levitin, cited in G.L. Vygodskaia and T.M. Lifanova, 'Lev Semenovich Vygotsky (Part 2)', *Journal of Russian and East European Psychology*, 37, 3 (1999), 3–80, p. 67.

[71] According to Marie Seton, Eisenstein first learned of Shereshevsky when Luria attended a party at the director's house in 1934 while Eisenstein was hosting the African American actor Paul Robeson in Moscow. Marie Seton, *Sergei M. Eisenstein: A Biography* (London: Dennis Dobson, 1978), p. 328. Eisenstein himself claims to have come across Shereshevsky earlier, saying that he spoke with Luria's patient in 1928 and 1933, S.M. Eisenstein, 'On Color' in *Selected Works, Volume 2, Towards a Theory of Montage*, ed. by Michael Glenny and Richard Taylor, trans. by Michael Glenny (London: BFI, 1991), p. 260.

[72] In January 1968 Luria wrote to Jerome Bruner to inform him that *Mind of a Mnemonist* would be published simultaneously in Russian and English, noting with incredulity that 105,000 copies had already been ordered in Russia. Bruner replied, joking that Luria was becoming as popular as 'Bulgakov or even Gorky!' See 11 January 1968, Luria to Bruner and 23 February 1968 Bruner to Luria. Cambridge, MA, Harvard University Archives (HUA), Jerome Bruner Papers, General Correspondence 1975–1977, HUG 4242.5, Box 88.

that the meanings of colours are deeply subjective and changeable: 'all these quests are individual and arbitrary.'[73] He celebrated the non-absolute relation, advocating the role of art in creating new modes of perception:

> a fixed and once-and-for-all scheme of mutually linked associations, would be profoundly inimical to the very nature of art. For one of the aims of art is to blaze new trails in our awareness of reality, to create new trains of association on the basis of utilising those which already exist.[74]

Despite insisting that existing correlations and associations are contingent, he does, however, imagine that they might eventually assume a fixed form (perhaps making a dutiful nod to the prevailing aesthetic discourses following the turn to socialist realism):

> in the progression of ever more perfect fusions of colour and sound, in the ever more perfect images that reflect the reality of our time, we shall also draw nearer to an ever fuller representation of the absolute truth of our unique, socialist way of life.[75]

New modes of perception will be blasted into consciousness but will ultimately settle to establish a harmonious truth. He implied that a fixed scheme of meaning would emerge under communism and that cinema could participate in bringing that future unity into being. Unity, again, was the ultimate goal.

Eisenstein may not have believed that synaesthetes could perceive essential connections between things. He did, however, claim that synaesthesia was a mode of perception that was closer to primitive forms of thought. Eisenstein described synaesthesia as 'sensuous primitive thinking', as an 'original bliss' that most humans had evolved away from.[76] He thus saw the synaesthete as being closer than most modern humans to the unity that he imagined characterised the earliest forms of perception. In Eisenstein's descriptions, Luria's patient replicated the ideal temporality

[73] Eisenstein, 'On Color', p. 258. For a detailed discussion of Eisenstein's engagement with synaesthesia, see Robert Robertson, *Eisenstein on the Audiovisual: The Montage of Music, Image and Sound in Cinema* (London: I.B. Tauris, 2009), pp. 140–201.

[74] Eisenstein, 'On Color', p. 260.

[75] Eisenstein, 'On Color', p. 261.

[76] Eisenstein, *Nonindifferent Nature*, trans. by Herbert Marshall (Cambridge: Cambridge University Press, 1987), p. 387.

he hoped to achieve through art; early forms of perception coexisted with advanced cognition:

> Comrade S's unique gifts were due to the fact that, while being a person of absolutely normal development, he had also retained into mature age all the characteristics of primal mental activity which other people lose as they develop intellectually and evolve towards normal, logical patterns of thought.[77]

Eisenstein was primarily interested in the sensuous associations that characterise synaesthesia. In the essay 'Vertical Montage' (1940), he described Shereshevsky's ability 'to see sounds as colours and hear colours as sounds', and to perceive vowels as 'gradations of light'.[78]

Luria's case history had a different emphasis: it dwelled on the emotional toll of his patient's vast memory:

> In this book I attempted to analyze the structure of the personality of a man who possessed an exceptional (in practice, unlimited) synesthetic memory. I concentrated not so much on the special features of his memory as on the consequences that this extraordinary memory had for the structure of the entire personality.[79]

Far from an almost magical capacity to perceive connections between things or an ability to apprehend some lost unity, Luria's narrative presents his patient's condition as distressing and alienating. As in his work with illiterate peasants and children, Luria insisted that the retention of more 'primitive' forms of thought could only be detrimental to future development; yet in this case the past was inescapable. The past (as memory) experienced by Shereshevsky was not harmony but an overwhelming chaos which alienated him from his immediate surroundings and interfered with his ability to maintain intimate connections with other people.

The condition described in *The Mind of a Mnemonist* contrasts starkly with *The Man With a Shattered World*. The latter's protagonist Zasetsky was cut off from his life history and sought desperately to reconnect to it—'he suffered intensely, and though his world had been devastated, in

[77] Eisenstein, 'Vertical Montage', *Selected Writings, Volume 2*, pp. 327–399, p. 368.

[78] Eisenstein, 'Vertical Montage', p. 368.

[79] Luria, 'A.R. Luria' in *A History of Psychology in Autobiography*, vol. 6, ed. by Gardner Lindzey (Cambridge, MA: Harvard University Press, 1974), pp. 253–292, p. 285.

the deepest sense he remained a man, struggling to regain what he had lost'[80]—whereas Shereshevsky suffered from *too much* connection to his previous experiences. His attachment to every past moment caused him to disconnect from the present. Even when he purposefully tried to rid himself of information—by scribbling his memories on scraps of paper and throwing them into the fire—he was ultimately incapable of forgetting anything. Sacks compared Luria's case history to the Jorge Luis Borges story 'Funes the Memorious', who

> knew by heart the forms of the southern clouds at dawn on the 30th of April, 1882, and could compare them in his memory with the mottled streaks on a book in Spanish binding he had only seen once and with the outlines of the foam raised by an oar in the Rio Negro the night before the Quebracho uprising.[81]

Sacks reflected on the complementarity of the two approaches—one literary, the other scientific—noting 'the unity between your "romantic science" and Borges' "scientific art"'.[82]

According to Luria's account, Shereshevsky retained words based on the sensual impression they made. As a child reading the Torah with no knowledge of Hebrew, he was able to retain long passages of text as the sounds registered as a series of splashes or puffs of steam that he could exactly reproduce in his mind. He created extraordinarily elaborate interior worlds in order to reconstruct things and all of these psychic fabrications were as imperishable as his memories of the external world. When presented with a mathematical equation he would narrativise it. Letters became people and symbols became actions: numbers might become jackdaws on a tree or cigarettes in a square. He would be able to reproduce a meaningless equation many years later by using this apparently circuitous visual method. Often he would memorise passages of text or long series of numbers by reconstructing a journey around Moscow, inserting objects along the route that corresponded to specific words. He describes placing a giant transparent head at Sretenski Gates and a statue of a Russian merchant woman on the pavement outside a movie theatre. He would then be able to repeat this 'journey' and reproduce all that he saw along the way,

[80] Luria, *Shattered*, p. 35.

[81] Jorge Luis Borges, 'Funes the Memorious' in *Ficciones*, trans. by Anthony Kerrigan (New York, NY: Grove Press, 2009), pp. 107–115, p. 112.

[82] Sacks to Luria, August 3, 1973, Oliver Sacks Archives.

occasionally stumbling if he had accidentally placed an object in an inconspicuous place: 'Sometimes I put a word in a dark place and have trouble seeing it as I go by.'[83] It was this associative capacity that inspired Eisenstein. He hoped to follow the example of Japanese Kabuki theatre, which he claimed succeeded in producing a synaesthetic effect on stage: 'we actually "hear movement" and "see sound".'[84]

Although Shereshevsky was able to recall the same passages of text or series of numbers many years later, his method for recalling was completely remote from received forms of logic. Luria describes it as 'a distinct type of dissociation'.[85] He discussed how Shereshevsky would employ the same laborious method of visualisation even when given a series of sequential numbers like 1,2,3,4: 'He had no need for logical organization, for the associations *his* images produced reconstituted themselves whenever he revived the original situation in which something had been registered in his memory.'[86] Similarly, Luria's patient found it difficult to follow stories as the narrative would be obscured by the rush of images associated with individual words or syllables. He was often unable to follow the meaning of a conversation as he would be distracted by the taste or smell produced by the person's voice. He noted that Vygotsky had a voice like crumbly yellow cheese, whereas Eisenstein seemed to speak with many voices at once, which was more like a bouquet: 'listening to him it was as though a flame with fibers protruding from it was advancing right toward me. I got so interested in his voice, I couldn't follow what he was saying.'[87]

Words for Shereshevsky sparkled or chafed. They could be bright or sharp, rough or bitter, prickly or sticky. The word 'soul' evoked an image of animals' livers and lungs on a table. At times his perceptions of words would conflict with their meanings, interfering with his daily activities. He describes his horror at discovering an ice cream seller with voice which evoked black coal. This overpowering sensual association prompted him to recoil with revulsion, rendering him incapable of purchasing any dessert. Similarly, if he read while eating the taste associations aroused by the food would blot out the meaning of the text. Although he learnt to direct or enhance his capacities through performing feats on stage, he was not

[83] Luria, *Mnemonist*, p. 33.

[84] Eisenstein, 'An Unexpected Juncture' in *Selected Writings, Volume 1*, pp. 115–122, p. 118.

[85] Luria, *Mnemonist*, p. 49.

[86] Luria, *Mnemonist*, p. 52.

[87] Luria, *Mnemonist*, p. 26.

able to fully control his capacity to recall; involuntary associations blotted out the voluntary. In conversation he would 'digress endlessly' distracted by his elaborate chains of association.[88] The words he encountered provoked such vivid visual associations that meaning was scattered, disrupting his ability to follow instructions or conversations: 'at each step he had to contend with superfluous images and sensations.'[89]

Eisenstein claimed that Shereshevsky's 'characteristics and abilities were retained alongside all the completely normal features of a fully developed mind and intellect',[90] whereas Luria stressed that his patient's synaesthesia precluded him from experiencing a 'normal' life. Even though Shereshevsky could work and perform basic routine tasks (unlike Zasetsky), he could not establish meaningful human relationships. He found it difficult to remember faces because he experienced them as 'changing patterns of light and shade'.[91] Shereshevsky's rich interior life overshadowed his connection to his immediate surroundings: '[at] times smoke or fog appears ... and the more people talk, the harder it gets, until I reach the point where I can't make anything out.'[92] Indeed, his imaginative capacity was so vivid that he could modify his body temperature or lower his threshold to pain simply by picturing himself in different surroundings. Luria characterised Shereshevsky's life as 'a struggle against images that kept rising to the surface of his mind'.[93] He dealt with the world, including his relationships with his closest family members, 'as though through a haze'.[94] Although he was more capable of functioning in society than Zasetsky, Luria presents Shereshevsky's case as the more tragic due to his patient's detachment and apathy. Though both men were focused on reaching a better future, only Zasetsky retained an acute awareness of the struggle

[88] Luria, *Mnemonist*, p. 116.
[89] Luria, *Mnemonist*, p. 104.
[90] Eisenstein, 'Vertical Montage', p. 368.
[91] Luria, *Mnemonist*, p. 53.
[92] Luria, *Mnemonist*, p. 35.
[93] Luria, *Mnemonist*, p. 87.
[94] Luria, *Mnemonist*, p. 118. Reed Johnson reveals fascinating discrepancies between Luria's case history and Shereshevsky's own versions of events recorded in personal notebooks. He also discusses Shereshevsky's biography in far more detail than in Luria's case history, revealing that Luria's patient had died of alcoholism in 1958, ten years before the case history was published. Johnson also challenges Luria's presentation of his patient's flawless memory, emphasising the effort and work it took Shereshevsky to remember, foregrounding instead the question of imagination. Reed Johnson, 'The Mystery of S. The Man with an Impossible Memory', *The New Yorker*, August 12, 2017, https://www.newyorker.com/books/page-turner/the-mystery-of-s-the-man-with-an-impossible-memory.

required to improve his condition. His relation to the future was rooted in an acknowledgment of the material conditions of the present he inhabited.

In contrast with Zasetsky, Luria claimed that Shereshevsky experienced his fantasy life as more 'real' than his everyday experiences: 'The boy was a dreamer whose fantasies were embodied in images that were all too vivid, constituting in themselves another world, one through which he transformed the experiences of everyday life.'[95] He often felt deep disappointment when reality failed to coincide with his imagined scenes. His everyday experiences thus faded into insignificance:

> He lived in wait of something that he assumed was to come his way, and gave himself up to dreaming and 'seeing' far more than to functioning in life. The sense he had that something particularly fine was about to happen remained with him throughout his life—something which would solve all his problems and make his life simple and clear. He 'saw' this and waited. … Thus everything he did in life was merely 'temporary', what he had to do until the expected would finally come to pass.
>
> I read a great deal and always identified myself with one of the heroes.[96]

This description of Shereshevsky's condition might also be applied to the progressive vision that undergirded the majority of Luria's publications, in which the present was implicitly treated as being less significant than the future it anticipated: the 'primitive' was encouraged to become 'civilised', the child to become an adult. The people Luria examined may have all inhabited the same historical moment but Luria placed some people further down a developmental ladder, as though they inhabited the past. Like Shereshevsky, who described himself as living always 'in the meantime', Luria's writings often relied on a temporality of deferral. He may not have sought to return to a lost origin point in the manner proposed by Eisenstein, but instead he idealised the imagined future: he analysed illiterate Uzbek people or children as exemplars of incipient forms of a linear trajectory of development with a predetermined end point. As such, he was not always attentive in recording or documenting the psychologies of those subjects as they existed in the present in his publications and did not always explore in detail the psychic impact of the specificities of their social or cultural contexts despite constantly claiming to do so.

In his work with illiterate peasants in the 1930s, Luria sought to replace the existing thought processes of those he was studying with more

[95] Luria, *Mnemonist*, p. 113.
[96] Luria, *Mnemonist*, p. 117.

'advanced' forms of thought. He wanted immediacy to be replaced by shared mediating structures. He framed this as a process of stabilisation, defined by the cumulative mastery of pre-existing social forms. His patients' psychic realities were thus subordinated to a fantasy of coherence located in an imagined future; an idealised mode of subjectivity was given precedence over the particular subjectivities he encountered. This progressive model of development had a counterpart in the imagined historical trajectory towards communism that permeated Soviet discourse, which similarly imagined spontaneity as a precursor to consciousness (to borrow once more the terms articulated by Lenin in *What is to be Done?*) and also relied on a temporality of 'in the meantime'.

The Russian title of *The Man with a Shattered World—A World Lost and Regained [Poteriannii i vozvrashchennii mir]*—gestures towards John Milton,[97] yet most of Luria's work, in sync with the Soviet experience, implied a paradise infinitely deferred. The socialist realist novel assures its readers that harmony will be reached, but this promised utopia was situated on an ever-receding horizon. Luria observed that when he went through transcriptions of conversations with Shereshevsky, he 'realized how difficult it was to single out what was essential from my conversations with S. from his endless digressions'.[98] Shereshevsky similarly perceived his own life as an endless digression, but digressions defined his condition, just as they constituted the real history of the Soviet Union. In both Shereshevsky's and Zasetsky's cases the ideal ending was never reached but that did not make the experiences that took place along the way any less significant or real.

Although Shereshevsky perceived himself as cut-off from the world, Luria's *The Mind of a Mnemonist*, by portraying the particularities of its protagonist's experiences, does situate the case in history. Eisenstein retreated from the present into the mists of prehistory, whereas Luria attended to the alienating haze that stood between patient and world as a real experience, rooted in the now. Like in *The Man with a Shattered World*, Luria did not occlude the struggles, detours, frustrations and feelings of disconnection from others that characterised his patient's condition. *The Man with the Shattered World* concludes by noting that despite the permanence of the damage inflicted to its patient-protagonist's brain, he will

[97] The Russian title of Luria's case history is *Poteriannii i vozvrashchennii mir* [*A Lost and Regained World*]. This recalls the Russian translations of John Milton's *Paradise Lost* [*Poteriannii rai*] and *Paradise Regained* [*Vozrashchennii rai*]. See Luria, *Romanticheskie Esse* [*Romantic Essays*] (Moscow: Pedagogika, 1996).

[98] Luria, *Mnemonist*, p. 116.

nonetheless continue striving to ameliorate his suffering, 'to try to recover what was irretrievable, to make something comprehensible out of all the bits and pieces that remain of his life'.[99] These 'romantic' case histories were informed by a methodological approach that succeeded in anchoring their subjects in their immediate contexts in a manner Luria's 'classical' works rarely achieved, even if this paradoxically involved describing anchorless, hazy, fragmented, alienated or temporally dislocated subjective states.

In her analysis of *Que Viva Mexico!* Salazkina describes how the sheer volume of 'raw filmic material' Eisenstein shot in Mexico was 'constantly overflowing and breaking out of the predetermined structure' he had mapped out.[100] In an analogous manner, the people Luria encountered in his clinical work sometimes exceeded the frameworks he sought to apply to them, but the 'romantic' case history eventually provided an adequate (if admittedly still mediated) form for capturing the particularities and vicissitudes of human consciousnesses, providing a suggestive model for psychological writing as a kind of art form in its own right. Sacks characterised Luria's project as reuniting science with art: 'the vision of such a poetic science, a total fusion of Art and Science, grips my heart and my mind with equal force, and tantalizes with glimpses of an old unity restored.'[101] Luria demonstrated that a finished work about an ongoing life could retain an unresolved, open-ended quality while leaving open the possibility for continual transformation.

Conclusion

In a letter to the Communist psychoanalyst Wilhelm Reich in 1934, Eisenstein attacked psychoanalysis for failing to situate neuroses in their social contexts and for thus affirming capitalist bourgeois society. He remarked with frustration that psychoanalysts were overly concerned with delineating norms which, he argued, served to uphold the existing state of things. Instead, he insisted that 'the boundary between the normal and the pathological cannot be drawn'.[102]

Luria's published works tended to be negatively structured around ideal norms of individual and social development and defined pathological

[99] Luria, *Shattered*, p. 158.

[100] Salazkina, p. 11.

[101] Sacks to Luria, June 14, 1975, Oliver Sacks Archives, Oliver Sacks Foundation, New York City.

[102] 'Sergei Eisenstein/Wilhelm Reich, Correspondence', ed. by Francois Albera, trans. by Ben Brewster, *Screen*, 22, 4 (1981), pp. 79–86, p. 85. Reich had visited the Soviet Union in 1929.

conditions in relation to norms from which they departed. His 'romantic' case histories, on the other hand, were portraits of pathologies that focused on sketching their protagonists without implicitly referring to a 'normal' figure. Without denying the suffering and difficulties his patients experienced, Luria's case histories nonetheless described experiences in such a way that they do not appear as deviations from a 'correct' path of development. The 'romantic' methodology Luria elaborated thus suggests an approach to human psychology that could open up possibilities for critiquing rather than affirming psychological norms but without romanticising mental illnesses or denying the lasting effects of brain injuries. Shifting the emphasis of psychology from an ideal vision of what is 'normal' to a consideration of the specificities of human subjectivities might then make it possible to interrogate the 'pathological social world[s]' people inhabit.[103] And this, perhaps, is where Luria's and Eisenstein's projects finally meet.

[103] 'Sergei Eisenstein/Wilhelm Reich, Correspondence', p. 85.

CHAPTER 7

Conclusion

Critical Theory and Romantic Science

In a letter to Jerome Bruner in 1958, Luria criticised what he saw as the most worrying trends then emerging in neurology:

> I think Engels was right by saying that "a function does not create an organ for its realisation", and that's not to do with metaphysics, that's a real biology, a real mode of formation of modes of living. It would be a big failure to deduce forms of behaviour from pre-existing forms of nervous cells and nervous connections.[1]

These remarks anticipated criticisms that have proliferated in the social sciences and humanities in the wake of the so-called decade of the brain in response to what Raymond Tallis has designated 'neuromania'.[2] Fernando Vidal, in a much-cited article that chimes with Luria's discussion with Bruner, claimed that the contemporary obsession with, and often blind faith in, neuroscientific research has led to a perceived identity between brain and self, giving birth to a new figure: the cerebral subject. Vidal,

[1] Alexander Luria to Jerome Bruner, August 15, 1958, Cambridge, MA, Harvard University Archives (HUA), Jerome Bruner Papers, General Correspondence 1975–1977, HUG 4242.5, Box 88.

[2] Raymond Tallis, *Aping Mankind: Neuromania, Darwinitis and the Misrepresentation of Humanity* (Durham, NC: Acumen, 2011), pp. 29–49. See also, ST Casper, *The Neurologists: A History of a Medical Speciality in Modern Britain, c. 1789–2000* (Manchester: Manchester University Press, 2014).

H. Proctor, *Psychologies in Revolution*, Mental Health in Historical Perspective, https://doi.org/10.1007/978-3-030-35028-4_7

along with his sometime co-author Francisco Ortega, seeks to unmask the social and political assumptions that allow 'neurocultures to prey on the most diverse pieces of evidence and the most varied beliefs in order to feed the ideology of brainhood'.[3]

Like Luria, Vidal challenges the identification of people with their brains and stresses the dehistoricising tendencies of much neuroscientific discourse, which tends to isolate brains from people and people from society. Vidal and Ortega argue that literary and cinematic representations of neurological conditions are capable of complicating the figure of the cerebral subject in their capacity to demonstrate that while people cannot live without their brains they are not therefore reducible to them.[4] They mention Luria's novelistic case histories as 'mirror image[s]'[5] of novels—including Richard Powers's *The Echo Maker* (2006) which takes its epigraph from an essay by Luria[6]—that are informed by neuroscientific perspectives both in their content and their narrative form. They identify the genre of the romantic case history as pioneered by Luria as 'an approach that attempts to recover the "I" or "who" of the patient's subjectivity from the "it" or "what" of the physical illness',[7] perceiving that in these works Luria succeeded in conveying subjective experiences of neurological disorders as part of people's unfolding lives in the world rather than as discrete, static phenomena.

[3] Fernando Vidal, 'Brainhood, Anthropological Figure of Modernity', *History of the Human Sciences*, 22, 1 (2009), 5–36, p. 20. See also Joelle M. Abi-Rached and Nikolas Rose in 'Birth of the Neuromolecular Gaze', *History of the Human Sciences*, 23, 1 (2010), 11–36 and *Neuro: The New Brain Sciences and the Management of the Mind* (Princeton, NJ: Princeton University Press, 2013). For a more explicitly left-wing critique of prevailing biological practices and the profits they generate, see Hilary and Stephen Rose, *Genes, Cells and Brains: The Promethean Promises of The New Biology* (London: Verso, 2012).

[4] Fernando Vidal and Francisco Ortega, *Being Brains: Making the Cerebral Subject* (New York, NY: Fordham University Press, 2017), p. 227.

[5] Vidal and Ortega, p. 200.

[6] On the significance of Powers's choice of epigraph as indicating a shared interest between novelist and neuroscientist in locating the soul 'within its social relations and material context', see Jenell Johnson, '"To find the soul, it is necessary to lose it": Neuroscience, Disability, and the Epigraph to The Echo Maker' in *Intersections: Essays on Richard Powers*, ed. by Stephen J. Burn and Peter Dempsey (Champaign, IL: Dalkey Archive Press, 2008), pp. 215–218, p. 217.

[7] Vidal and Ortega, p. 201. See also G. Thomas Couser's discussion of Oliver Sacks in *Vulnerable Subjects: Ethics and Life Writing* (Ithaca, NY: Cornell University Press, 2004), pp. 74–122.

'Critical Neuroscience' is a recent endeavour informed by the work of the Frankfurt School, whose concerns overlap with Vidal and Ortega's. This emerging body of criticism attempts to probe the social and political implications of neuroscientific research: the values it mobilises, the interests it serves, the vision of nature and subjectivity it participates in construing. In their introduction to an interdisciplinary collection of essays inaugurating this project, Suparna Choudhury and Jan Slaby state: 'The goal is to work towards an integrated approach to behaviour that situates the brain and cognition in the body, the social milieu and the political world.'[8] The essays in the collection explore the 'reciprocal interactions between neuroscience and social life' and are committed to a 'politics of situated subjectivity', which refuses to accept the identity of brain and self or to cordon brains off from society or history.[9]

When I began this project I was struck that Luria often described the brain in a manner amenable to these descriptions. I therefore set out to analyse his work as if he were a kind of proto-critical neuroscientist. But my attempts to do so were soon confounded when I began trying to situate his arguments about the historical situatedness of subjectivity in their own history. Although, as Powers's *The Echo Maker* demonstrates, Luria's statements about the relationships between the social and the psychological or neurological make appealing epigraphs, plucking sentences out of Luria's publications without considering the specific historical situations in which the people he analysed lived risks obscuring the very factors to which such statements proclaim to be alert.

In his contribution to the collection *Critical Neuroscience* Martin Hartmann declares: 'There is as yet no critical theory of the neurosciences'.[10] Hartmann's piece is the only essay in the collection which explicitly attempts to outline what such a theory might entail. He envisages an approach that would engage directly with contemporary neuroscientific research in order to lay bare its coercive aspects. In setting out this potential model, he draws primarily on the work of Jürgen Habermas who, he argues, built on the early foundations laid by Max Horkheimer in the

[8] Suparna Choudhury and Jan Slaby, 'Introduction' in *Critical Neuroscience: A Handbook of the Social and Cultural Contexts of Neuroscience*, ed. by Choudhury and Slaby (Chichester: Wiley-Blackwell, 2012), pp. 1–26, p. 3.

[9] Choudhury and Slaby, p. 7, p. 10.

[10] Martin Hartmann, 'Against First Nature: Critical Theory and Neuroscience' in *Critical Neuroscience*, pp. 67–84, p. 68.

1930s but 'dropped this Marxist frame'.[11] Unlike Hartmann, I embarked on this project with the intention of holding on to Marxist frames and was particularly keen to emphasise the distinctly Marxist aspects of Luria's work. I hoped *Psychologies in Revolution* could participate in challenging enmeshed economic and scientific structures in the present, identified by Hartmann as a central element of critical theoretical approaches, by excavating an explicitly Marxist theory of subjectivity from the past. In *The New Spirit of Capitalism* (2007) sociologists Luc Boltanski and Eve Chiapello discuss how descriptions of the brain recall descriptions of the free market economy, observing that such vocabularies function to naturalise the current economic system, making it seem impervious to change, and thus undermine people's capacity to protest, resist or struggle against it.[12] Though wary of seeming to validate 'actually existing socialism' I was nonetheless interested in exploring whether a psychological and neuroscientific theory conceptualised explicitly in opposition to capitalism and developed in a non-capitalist context might provide a way out of this impasse.[13]

This proved a more fraught endeavour than I first envisioned, however. While some of the theoretical underpinnings of Luria's work had their roots in Marxism and were specific to his Soviet context focusing on those influences risked downplaying others.[14] Moreover, the more of his work I read the more I felt that it would be disingenuous, indeed perverse, to focus on the theoretical underpinnings of his work at the expense of the historical circumstances that shaped his research and the lives of the people he encountered while conducting it. The example of post-Soviet Russia also serves to complicate characterisations of the ways current scientific paradigms are shaped by capitalism (while simultaneously seeming to

[11] Hartmann, p. 71.

[12] Luc Boltanski and Eve Chiapello, *The New Spirit of Capitalism* (London: Verso, 2007), pp. 149–150.

[13] The question could arguably be posed the opposite way around as well: that an obsession with the materiality of the brain evident in dominant Soviet scientific paradigms actually *prefigured* the Western 'age of the brain' and so calls into question its distinctly capitalist qualities.

[14] Though, of course, Marxism is by no means homogeneous, static and internally coherent, the nuanced and idiosyncratic Marxism of Vygotsky was distinct from the orthodox Marxist-Leninism of official Soviet discourse and Luria drew on both (alongside an eclectic set of influences from 'bourgeois' Western psychology). On the question of how 'communist' psychiatry under state socialism was, see Gregory Eghigian, 'Was There a Communist Psychiatry? Politics and East German Psychiatric Care, 1945–1989', *Harvard Review of Psychiatry*, 10, 6 (2002), 364–368.

function as perfect metaphors for it), underlining the necessity to emphasise historical particularity. In *Shock Therapy: Psychology, Precarity and Postsocialist Russia* (2018), anthropologist Tomas Matza analyses the surge of interest in an eclectic range of psychotherapeutic practices in the wake of the collapse of the Soviet Union in 1991, arguing that while this could be seen as a generalised 'symptom of neoliberal capitalism's arrival', through his ethnographic work he found that it also, more crucially, responded to 'the ending of a way of life', to uncertainty about the future and to material hardships in the present:[15]

> what was at stake for both men and women caught up in the psychotherapeutic turn was not so much the construction of the deep psychological self that scholars term *neoliberal subjectivity*, but a search for modes of truth telling about experience, emotional harm, or violence, and a pursuit of sociality in the privatized spaces of postsocialism.[16]

The transition to the market economy prompted new engagements with a range of psy-ences and new discourses about the self, yet these phenomena are not best approached through analysing abstract representations of the market and models of the mind that overlap with those representations, but in thinking through ways people negotiated dramatic shifts in their daily lives under the newly introduced economic and political regime. Capital, after all, is a social relation.[17]

Although I felt more at home writing about theories of historical materialism than about historical materials, I nonetheless became convinced that this project demanded I attempt to write as a historian rather than as a critical theorist, so I attempted to imagine history writing as a mode of critical theory. Though not in the way I initially conceived, in concluding my account of Luria's 'romantic' science, I want to propose that *Psychologies in Revolution* might still be read as a contribution to critical neuroscience

[15] Tomas Matza, *Shock Therapy: Psychology, Precarity, and Well-Being in Postsocialist Russia* (Durham, NC: Duke University Press, 2018), pp. 3–4.

[16] Matza, p. 8. See also: Svetlana Alexievich, *Second-Hand Time*, trans. by Bela Shayevich (London: Fitzcarraldo, 2016), Serguei Alex Oushakine, *The Patriotism of Despair: Nation, War, and Loss in Russia* (Ithaca, NY: Cornell University Press, 2009) and Eugene Raikhel, *Governing Habits: Treating Alcoholism in the Post-Soviet Clinic* (Ithaca, NY: Cornell University Press, 2016).

[17] On tensions between representations and experiences of capitalism and the problem of representing a totality, see Alberto Toscano and Jeff Kinkle, *Cartographies of the Absolute* (Winchester: Zero Books, 2015).

by demonstrating the cultural and social contingency of both psychiatric and neuroscientific knowledge and the people whose brains or minds it seeks to understand, denaturalising the present through a rehistoricisation of psy-entific theories.[18] Moreover, I propose that Luria's 'romantic' science—not an abstract theoretical model but a form of writing attentive to the realities, specificities and complexities of subjective experience—offers a model for approaching (human) nature that goes beyond the characterisations of science outlined in canonical texts of critical theory.

* * *

In the programmatic essay 'Traditional and Critical Theory' (1937), Max Horkheimer argued that traditional theory naturalises the social by imagining a world driven by alien external forces beyond the control of human agency. Traditional theory, Horkheimer claims, serves to uphold the existing political order: 'The scholar and his science are incorporated into the apparatus of society; his achievements are a factor in the conservation and continuous renewal of the existing state of affairs, no matter what fine names he gives to what he does.'[19] A critical theorist, on the other hand, is 'wholly distrustful of the rules of conduct with which society as presently constituted provides each of its members'.[20] In the aphorism 'Ego is id' in *Minima Moralia* (1951), in an argument that recalls Horkheimer's earlier attack on traditional theory, Theodor Adorno suggested that the discipline of psychology is structured to uphold the status quo, identifying the concomitant emergence of psychology with the 'rise of the bourgeois individual'.[21] Adorno conceived of psychology as a discipline that is incorporated into the apparatus of bourgeois society. Psychology, he claimed, is

[18] On history as a means of combatting the dominance of neuroscientific paradigms for understanding the world (rather than on histories of neuroscience specifically), see Roger Smith, 'Resisting Neurosciences and Sustaining History', *History of the Human Sciences*, 32, 1 (2019), 9–22. Historicising neuroscience also indicates the limits of using current scientific or psychiatric categories for analysing the actions and experiences of people from the past, an approach evident in, for example, Lynn Hunt, 'The Experience of Revolution', *French Historical Studies*, 32, 4 (2009), 671–678.

[19] Max Horkheimer, 'Traditional and Critical Theory' in *Critical Theory: Selected Essays* (New York, NY: Continuum, 1972), pp. 188–243, p. 196.

[20] Horkheimer, p. 207.

[21] Theodor Adorno, *Minima Moralia: Reflections on a Damaged Life*, trans. by E.F.N. Jephcott (London: Verso, 2005), p. 63.

premised on rationalisation; it elevates human subjects above all else but in so doing negates subjectivity and thus transforms human beings into 'absolute object[s]'.[22] It therefore participates in the principles of human domination and exchangeability inherent to capitalism. By treating the psychic life of suffering people as an interior space to be externally controlled, the logic of property ownership is replicated in the therapeutic encounter. Adorno attacked psychoanalysis specifically for its equation of healing with adjustment to the existing state of things: patients are enjoined to align themselves with the reality principle, which only serves to uphold alienation.

In emphasising the relationship between human subjects and their historical circumstances, and by asserting that humans could actively intervene in history, Luria explicitly sought to differentiate his approach from the Western bourgeois traditions attacked by Horkheimer and Adorno. Ultimately, however, (and this is where my project departed from what I had initially envisioned writing) Luria was not able to evade the affirmative relationship of psychology to existing reality identified by Horkheimer in traditional theory and Adorno in psychoanalysis. Martin Jay defines Western Marxism as a diffuse and undogmatic group of intellectuals open to non-Marxist schools of thinking and critical of official Communist Party positions, and includes Adorno and Horkheimer in this canon of thinkers. He frames Western Marxism as the 'negation of its official Soviet (or Eastern) counterpart. The latter has been turned into a doctrinaire ideology of legitimation by a tyrannical regime, whereas Western Marxism, nowhere in power, had retained the libertarian, emancipatory hopes of the socialist tradition.'[23] *Psychologies in Revolution* says more about the messy and often traumatic realities of actually existing socialism than it does about 'Eastern' Marxist theory in the abstract, however. Historical antagonisms and forms of ongoing social injustice were often overlooked by Luria in theory but still documented in practice. As is evident in *The Man with a Shattered World*, even when attempting to narrate his experiences in accordance with an Eastern Marxist teleological conception of historical

[22] Adorno, *Minima Moralia*, p. 63.

[23] Martin Jay, *Marxism and Totality: The Adventures of a Concept from Lukács to Habermas* (Berkeley, CA: University of California Press, 1984), p. 2. In his discussion of Western Marxism, Maurice Merleau-Ponty describes Soviet Marxism as the substitution of a 'simple schema of progress for the difficult reading of the anticipations and delays of history and for the rigorous examination of revolutionary society'. Maurice Merleau-Ponty, *Adventures of the Dialectic*, trans. by Joseph Bien (London: Heinemann, 1974), p. 71.

and individual development, Zasetsky's neurological condition pulled against this characterisation.

Drawing inspiration from art historians and cultural theorists who returned to utopian post-revolutionary artistic projects in the hope of excavating something of their hopeful world-transforming utopian visions for the present, when I embarked on this project I imagined that aspects of Luria's explicitly left-wing approach to human subjectivity might be capable of challenging prevailing ideologies of brainhood under capitalism today.[24] I speculated that Luria's publications might provide the basis for an immanent critique of the scientific paradigms his work is still included in, like a kind of Trojan Horse sitting quietly within the city walls. I sprinkled my writing with citations from the work of Walter Benjamin who declared that the historical materialist approaches the past in order to surprise or disrupt the present. History writing for Benjamin is a political endeavour addressed to now:

> [The] dialectical penetration and actualisation of former contexts puts the truth of all present action to the test. Or rather, it serves to ignite the explosive materials latent in what has been. ... To approach, in this way, 'what has been' means to treat it not historiographically, as heretofore, but politically, in political categories.[25]

The historical materialist fans 'the spark of hope in the past', to provoke a bright but destructive conflagration in the present.[26] I identified a Benjaminian spark in Yuri Olesha's novel *Envy* (1927), viewing his

[24] I discuss the paradoxes of this mode of revolutionary excavation in Hannah Proctor, 'Revolutionary Commemoration', *Radical Philosophy*, 2, 1 (2018), 47–57. Works that I read with excitement in the early stages of this project and which I initially hoped my work on Luria might draw inspiration from include: Susan Buck-Morss, *Dreamworld and Catastrophe: The Passing of Mass Utopia in East and West* (Cambridge, MA: MIT Press, 2000), T.J. Clark, *Farewell to an Idea: Episodes in the History of Modernism* (New Haven, CT: Yale University Press, 1999), Christina Kiaer, *Imagine No Possessions: The Socialist Objects of Russian Constructivism* (Cambridge, MA: MIT Press, 2008) and Richard Stites, *Revolutionary Dreams: Utopian Vision and Experimental Life in the Russian Revolution* (Oxford: Oxford University Press, 1989).

[25] Walter Benjamin, *The Arcades Project*, ed. and trans. by Rolf Tiedemann (Cambridge, MA: Belknap Press, 1999), p. 392.

[26] Benjamin, 'Theses on the Philosophy of History' in *Walter Benjamin: Selected Writings, Vol. 4, 1938–1940*, ed. by Howard Eiland and Michael W. Jennings, trans. by Edmund Jephcott (Cambridge, MA: Belknap Press of Harvard University Press, 2006), pp. 389–399, p. 391.

metaphor for bringing the past into illuminating if fleeting dialogue with the present as a guiding light for my own work:

> You know, it happens that an electric bulb will unexpectedly go out. It burned out, you say. And if you shake this burned out bulb then it'll light up again and will still burn for some time. Inside the bulb a catastrophe is taking place. Tungsten filaments break off and through the contact of the fragments life returns to the bulb. A short, unnatural, undisguisedly doomed life—a fever, an overly bright incandescence, a brilliance. Then will come darkness, life will not return, and in the darkness the dead scorched filaments will only rattle. Do you understand? But the brief brilliance is beautiful. I want to shake the heart of the burned out epoch. The heart-bulb, so that the fragments would touch and to bring about a momentary, beautiful brilliance.[27]

Ultimately, however, I did not feel able to present Luria's work as an eruptive and disruptive collision capable of unsettling or igniting the oppressive present in good faith. I found myself dwelling on moments of darkness rather than incandescence. I found myself more interested in exploring Luria's works written in a modest quotidian register than in those laced with bombastic revolutionary proclamations (and I was forced to acknowledge that many of his texts contained no such proclamations at all). I found explosive materials but in the form of shrapnel fragments and detonated bombs from a war of unprecedented violence and destruction. Yet it was precisely here, in the aftermath of that global catastrophe, that I did find a redemptive and politically hopeful spark in Luria's work. Unlike in Luria's most explicitly Marxist theoretical writings published in the aftermath of the October Revolution or during the period of the First Five Year Plan, *The Man With a Shattered World* is not characterised by thrilling flashes of breathless utopianism proclaiming the dawn of a new emancipated society but instead describes an interminable and arduous yet determined and hopeful process of recovery and repair. In *Enduring Time* (2017) Lisa Baraitser attends to practices of endurance, maintenance and care, observing that 'Maintenance is not the time of generation or production, or the eruption of the new. It is not revolutionary time, but the lateral time of "on-go" that tries to sustain an elongated present.'[28] Such practices and the drawn-out temporalities they imply may lack the seductive

[27] Yuri Olesha, *Envy*, trans. by T.S. Berczynski (London: Ardis, 1975), p. 71.

[28] Lisa Baraitser, *Enduring Time* (London: Bloomsbury, 2017), p. 52.

vibrancy of revolutionary rupture but, she argues, they are not therefore apolitical. Indeed, Luria's work demonstrates that moments of dramatic change are accompanied and followed by quieter processes that, however slow, might still be transformative.

* * *

In 'Reconciliation Under Duress' (1958), a rare example of an essay in which he explicitly attacked official Communist Party orthodoxy, Adorno articulated the paradoxical relationship between Marxist rhetoric and Soviet reality at stake in my discussions of Luria. The essay is a searing critique of the Hungarian philosopher György Lukács, who remained loyal to the Communist Party throughout the Stalin era. Adorno claims that 'under the mantle of an ostensibly radical critique of society', Soviet Marxism had 'degraded the philosophy it proclaimed to the level of a mere instrument in the service of its rule'.[29] According to Adorno, 'officially licensed' Marxists like Lukács proclaimed their work dialectical, while producing thoroughly undialectical texts, which denied the persistence of oppression in Soviet society.[30] Adorno insisted that in the Soviet Union the antagonisms and inequalities the revolution promised to sweep away continued but were disavowed by the officially sanctioned rhetoric: 'the cleavage, the antagonism persists, and it is a sheer lie to assert that it has been "overcome", as they call it, in the states of the Eastern bloc.'[31] Adorno perceived to borrow Susan Buck-Morss's formulation, 'that actual past history was not identical to the concept of history (as rational progress)'.[32] In the case of the Soviet Union, however, he did not devote much energy to probing the implications of the tensions between the two.

Psychologies in Revolution, more sympathetic to Luria than Adorno was to (late) Lukács, has attempted to discern and describe tensions between duress and reconciliation evident in Luria's works; it identified moments in which theories of historical materialism collided with materials of history (i.e. moments in which Luria's understanding of individual development as a parallel to historical teleology was complicated by his empirical

[29] Theodor Adorno, 'Reconciliation Under Duress' in *Aesthetics and Politics*, ed. by Frederic Jameson (London: Verso, 2007), pp. 165–195, p. 166.

[30] Adorno, 'Reconciliation…', p. 166.

[31] Adorno, 'Reconciliation…', p. 195.

[32] Susan Buck Morss, *The Origin of Negative Dialectics: Theodor W. Adorno, Walter Benjamin and the Frankfurt Institute* (Hassocks, NJ: Harvester Press, 1977), p. 49.

findings, as was the case with the experiments conducted in Central Asia). Horkheimer noted that the identification of people with a 'critical mind with their society is marked by tension, and the tension characterises all the concepts of the critical way of thinking'.[33] Although Luria's work tended to present a vision tacitly in harmony with his society and often affirmed harmony as an ideal goal of both individual and historical development, reading his publications closely and critically revealed that tensions nonetheless persisted. Horkheimer's distinction between traditional and critical theory might be aligned with Luria's distinction between classical and romantic science. By describing lived realities in the present Luria's romantic case histories succeeded in bringing to the fore the kinds of tensions that characterise critical theory, and even, I want to conclude by tentatively venturing, going beyond it.

* * *

In 'Primitive Man and His Behaviour' in *Studies in Behaviour: Ape, Primitive and Child* (1930), Luria and Vygotsky argued that the use of magic in primitive societies demonstrates humanity's desire to control nature and itself, and thus represents a significant progressive step towards civilisation. The authors point to magic's rational kernel, arguing that casting spells to create rain or love are embryonic forms of civilisation: 'we see in magic in its undeveloped form both the future technique for controlling nature and the cultural techniques for the control of man's own behaviour.'[34] The husk of irrationality is discarded when humanity develops more advanced techniques for mastering its environment. Luria and Vygotsky's argument is almost the inverse of that put forth by Adorno and Horkheimer in *The Dialectic of Enlightenment*, in which enlightened rationality emerges through the repression or disavowal of irrationality, rather than through its extirpation. Adorno and Horkheimer's argument agrees with Vygotsky and Luria's insofar as they conceive of domination over nature as a central tenet of magic. However, Luria and Vygotsky's argument conforms with Adorno and Horkheimer's definition of Enlightenment thinking as they champion disenchantment and scientific progress. Indeed,

[33] Horkheimer, p. 208.

[34] Luria and Vygotsky, 'Primitive Man and His Behaviour' in *Studies on the History of Behaviour: Ape, Primitive and Child*, ed. and trans. by Victor I. Golod and Jane E. Know (Hillsdale, NJ: Lawrence Erlbaum, 1993), pp. 79–139, p. 134.

Adorno and Horkheimer explicitly argue that state socialism, by clinging to the heritage of bourgeois philosophy, failed to escape the dialectic of Enlightenment.[35] For Adorno and Horkheimer, although magic itself already entails enlightened thinking, it retains a sensitivity to specificity which modern science lacks. The 'corrosive rationality'[36] of Enlightenment obliterates the differences between things: 'All gods and qualities must be destroyed'.[37] This 'amputation of the incommensurable',[38] which reduces nature to exchangeable standardised chunks, is achieved through abstraction: 'Bourgeois society is ruled by equivalence. It makes dissimilar things comparable by reducing them to abstract quantities.'[39] For Adorno and Horkheimer, humans in advanced industrial societies, like terrified rabbits dissected in laboratories, are reduced to isolated but identical samples of their species. Concatenated nature is split into discrete specimens of matter, individual suffering is elided and the thing-like bourgeois subject emerges.

Yet, they claim, the irrational cannot be dispensed with so easily. Indeed, proclaiming its disappearance paradoxically functions to increase its strength: 'the curse of irresistible progress is irresistible regression.'[40] Instead they argue—dialectically—that humanity's increased power over nature is coupled with the increased subjugation of humanity. By eviscerating the particular, the Enlightenment left the 'uncomprehended whole'[41] (nature) free to rebound on the human subjects who had attempted to master and control it. This violence, they claim, has its origins in abstraction which, they argue, is painfully concrete.

One goal of Luria's research was to identify, trace and encourage the emergence of the capacity for abstract thought, defined as the capacity to generalise, categorise and classify objects according to pre-existing socially established criteria. He viewed this as crucial to the progress of civilisation. Following Adorno and Horkheimer's argument, Luria's faith in the emancipatory potentials of abstract thinking overlooked the structure of

[35] Theodor Adorno and Max Horkheimer, *The Dialectic of Enlightenment: Philosophical Fragments*, ed. by Gunzelin Schmid Norr, trans. by Edmund Jephcott (Stanford, CA: Stanford University Press, 2002), p. 32.

[36] Adorno and Horkheimer, p. 4.

[37] Adorno and Horkheimer, p. 5.

[38] Adorno and Horkheimer, p. 9.

[39] Adorno and Horkheimer, p. 4.

[40] Adorno and Horkheimer, p. 28.

[41] Adorno and Horkheimer, p. 33.

domination inhering within it. However, as was evident in his experiments in Central Asia, a contradiction existed between Luria's discussions of abstraction as a cognitive capacity, on the one hand, and as a research methodology, on the other. For Luria not only advocated abstraction as a mode of thought but simultaneously insisted that a properly Marxist approach to psychology should treat cognition as time bound and historical. His scientific approach, particularly his definition of romantic science, claimed to retain a sensitivity to the specificities of his objects of analysis that seem alert to the rationalising tendencies of scientific abstraction condemned by Adorno and Horkheimer. Luria described illiterate people in Uzbekistan as being insufficiently advanced due to the 'unmediated' nature of their thought processes, yet immediacy was paradoxically what he strove for when he constructed his experiments;[42] Luria extolled the capacity for abstraction whilst eschewing it as a research methodology (in theory if not always in practice).

In his descriptions of 'classical' science, Luria characterises abstraction as a withdrawal from reality, which drains qualities from the concrete. In these discussions he draws close to Adorno and Horkheimer's arguments and, as I proposed above, his distinction between classical and romantic science recalls the distinction between traditional and critical theory Horkheimer outlined in his 1930 essay. Similarly, in *Negative Dialectics*, Adorno advocated a mode of engagement with the world that resists abstraction by attending to objects in their specificity, which seems to chime with Luria's definition of romantic science: 'To yield to the object means to do justice to the object's qualitative moments.'[43]

In the context of the Central Asian expeditions, Luria's declared sensitivity to cultural particularity was complicated by his continued imposition of pre-existing normative frameworks onto the Uzbek people he encountered, muffling the specificities of their utterances. Classical scientific abstraction returned through the back door. Ultimately, however, as the example of his descriptions of discussions with *ichkari* women demonstrates, Luria's publications based on his research in Central Asia did not succeed in presenting the people he described as identical with his frameworks, attesting instead to continued tensions between concrete and abstract, romantic and classical.

[42] Luria, *Cognitive Development*, p. 30.

[43] Adorno, *Negative Dialectics*, trans. by E.B. Ashton (New York, NY: Seabird Press, 1973), p. 43.

The two-fold character of Luria's understanding of abstraction, which was simultaneously 'classical' and 'romantic', a binary Sacks defined in a letter to Luria as one 'between the particularity of the experience (to which I must minutely and faithfully adhere) and the universality of concepts (to which I ardently, and perhaps dangerously, aspire)',[44] recalls the two-fold character of the dialectic of Enlightenment in which domination and emancipation are constantly intertwined. A dialectical understanding of the relation between abstract and concrete is acknowledged by Luria when, in his discussion of scientific methodologies, he invokes Marx's 'strange-sounding expression "ascending to the concrete"'.[45] Similarly, the properly dialectical structure of Adorno and Horkheimer's argument ensures that qualities are never successfully liquidated by abstraction but remain as a substratum of all quantities.[46] After all, they did not advocate returning to some lost past devoid of abstraction (indeed, they are clear that such a past never existed as 'myth is already Enlightenment').[47] Splintered shards of hope glimmer faintly in their austere prose. Ultimately they insist that 'every form of coercive power' tends towards a 'negative utopia'.[48] And this insight is applicable to Luria's project.

The closing chapters of this book argued that Luria's 'romantic' case histories succeeded in prioritising the analysis of his subjects over his frames for understanding them, attending to rather than seeking to banish the haze or 'troubled air' (to repeat Fred Moten's phrase), separating him from them or them from their own surroundings. As such, Luria's attempt to develop a scientific method attentive to history might itself challenge Adorno and Horkheimer's sweeping characterisation of science as necessarily on the side of manipulation and unification, in favour of a dialectical approach to objects—or, rather, to human subjects—that does not reduce them to mere quantities. In *The Concept of Nature in Marx* (a book which emerged from a PhD supervised by Adorno and Horkheimer), Alfred Schmidt argues that science as such 'is undialectical in that it does not

[44] Sacks to Luria, July 14, 1974, Oliver Sacks Archives, Oliver Sacks Foundation, New York City.

[45] Luria, *Making of Mind*, p. 178. This question is explored in detail in the work of the Soviet philosopher E. Ilyenkov (who collaborated with psychologists associated with Luria in Moscow in the post-war period). See E.V. Ilyenkov, *The Dialectics of the Abstract and Concrete in Marx's Capital*, trans. by Sergei Syrovatkin (Delhi: Aakar, 2008).

[46] Adorno, *Negative Dialectics*, p. 43.

[47] Adorno and Horkheimer, p. xviii.

[48] Adorno and Horkheimer, p. 54.

reflect the historical mediation of its objects'.[49] Luria's case histories suggest that this particular line of 'Western Marxist' thought is overly pessimistic for its refusal to consider the possibility of an approach to nature that might succeed in being attentive to history. The romantic science, which Luria inaugurated in his case histories, suggests that scientists might be capable of producing work that refuses to deny the historicity of nature and thus call into question the natural-seeming social.[50]

[49] Alfred Schmidt, *On the Concept of Nature in Marx* (London: New Left Books, 1971), p. 55.

[50] Such a vision is articulated by Adorno in his 1932 essay 'The Idea of Natural History', *Telos*, 60 (1984), 111–124. A similar conclusion is reached by John Bellamy Foster in *Marx's Ecology: Materialism and Nature* (New York, NY: Monthly Review Press, 2000).

Index[1]

A

B

C

[1] Note: Page numbers followed by 'n' refer to notes.

H. Proctor, *Psychologies in Revolution*, Mental Health in Historical Perspective, https://doi.org/10.1007/978-3-030-35028-4

H

I

J

K

L

M

The manufacturer's authorised representative in the EU is Springer Nature Customer Service Centre GmbH, Europaplatz 3, 69115 Heidelberg, Germany. If you have any concerns regarding our products, please contact ProductSafety@springernature.com

Printed and bound by CPI Group (UK) Ltd, Croydon, CR0 4YY
05/10/2026
02228143-0003